STUDENT RESOURCE MANUAL
to accompany

Elementary Algebra
sixth edition

by Charles P. McKeague

JOHN GARLOW
Tarrant County College -Southeast Campus

LORI PALMER
Utah Valley State College

Saunders College Publishing
A Division of Harcourt College Publishers

Fort Worth Philadelphia San Diego New York Orlando Austin
San Antonio Toronto Montreal London Sydney Tokyo

Copyright © 2000, 1995, 1990, 1987 by Harcourt, Inc.

All rights reserved. No part of this publication may be reproduced or transmitted in any form or by any means, electronic or mechanical, including photocopy, recording, or any information storage and retrieval system, without permission in writing from the publisher.

Requests for permission to make copies of any part of the work should be mailed to the following address: Permissions Department, Harcourt, Inc., 6277 Sea Harbor Drive, Orlando, Florida 32887-6777.

Printed in the United States of America

ISBN 0-03-026284-4

012 202 765432

Table of Contents

Preface	i.

Part One: Solutions

Chapter 1: The Basics	1
Chapter 2: Linear Equations and Inequalities	14
Chapter 3: Linear Equations and Inequalities in Two Variables	55
Chapter 4: Systems of Linear Equations	75
Chapter 5: Exponents and Polynomials	96
Chapter 6: Factoring	120
Chapter 7: Rational Expressions	142
Chapter 8: Roots and Radicals	175
Chapter 9: More Quadratic Equations	195

Part Two: Math in Practice Problems

Introduction	225
Math In Practice Problems	A1-A107
Answers to Math In Practice Problems	A109-118

Multimedia CD-ROM Single User License Agreement	B-1
System Requirements	B-2
Installation Instructions	B-2
Technical Support	B-2

is

A Harcourt Higher Learning Company

Now you will find Saunders College Publishing's distinguished innovation, leadership, and support under a different name . . . a new brand that continues our unsurpassed quality, service, and commitment to education.

We are combining the strengths of our college imprints into one worldwide brand: Harcourt

Our mission is to make learning accessible to anyone, anywhere, anytime—reinforcing our commitment to lifelong learning.

We are now Harcourt College Publishers. Ask for us by name.

One Company
"Where Learning Comes to Life."

www.harcourtcollege.com
www.harcourt.com

Preface

This manual is comprised of two parts.

Part One of the manual contains complete solutions to every odd-numbered problem in the problem sets, chapter review sets, and the cumulative reviews. Additionally, it contains solutions to every problem in the chapter tests. These solutions were completed by Dr. John Garlow from Tarrant County College – Southeast Campus.

This manual should be used as a reference for attempting the solutions on your own. The importance of your initial effort can not be overstated. The quality of your learning experience in mathematics is directly related to the effort you apply in working the problems assigned to you.

John would like to that James LaPointe of Harcourt College Publishers for his editorial assistance and for giving him the opportunity to author this manual. He also thanks Charles P. McKeague for writing the textbook and carefully revising it to make the writing of the solutions manual go more smoothly. Heartfelt thanks go to his wife Suzanne Garlow for the keyboarding, proof reading and final formatting. He is also appreciative of Sudhir Goel of Valdosta State University for his thoughtful and careful accuracy checking of the initial manuscript.

Part Two of the manual begins on page 223 and contains the printed versions of the problems that are found on the *Math in Practice: An Applied Video Companion* CD-ROM. This CD-ROM is inserted on the inside back cover of this manual. The pages are formatted in such a way to allow room for you to complete the problems and turn them in to your professor. The CD-ROM was created by Utah Valley State College's Lori Palmer and is designed to show students that precalculus concepts arise in real life. Lori has conducted over 20 engaging interviews with professionals in such fields as aviation, food services, banking, and environmental sciences. Each video clip is accompanied by two problems. Answers to the problems are presented at the back of the manual.

Part One: Solutions

SOLUTIONS TO SELECTED PROBLEMS

Chapter 1
The Basics

1. $x + 5 = 14$
3. $5y < 30$
5. $3y \le y + 6$
7. $\frac{x}{3} = x + 2$
9. $3^2 = 3 \cdot 3 = 9$
11. $7^2 = 7 \cdot 7 = 49$
13. $2^3 = 2 \cdot 2 \cdot 2 = 8$
15. $4^3 = 4 \cdot 4 \cdot 4 = 64$
17. $2^4 = 2 \cdot 2 \cdot 2 \cdot 2 = 16$
19. $10^2 = 10 \cdot 10 = 100$
21. $11^2 = 11 \cdot 11 = 121$
23. $2 \cdot 3 + 5 = 6 + 5 = 11$
25. $2(3 + 5) = 2(8) = 16$
27. $5 + 2 \cdot 6 = 5 + 12 = 17$
29. $(5 + 2) \cdot 6 = 7 \cdot 6 = 42$
31. $5 \cdot 4 + 5 \cdot 2 = 20 + 10 = 30$
33. $5(4 + 2) = 5(6) = 30$
35. $8 + 2(5 + 3) = 8 + 2(8) = 8 + 16 = 24$
37. $(8 + 2)(5 + 3) = 10(8) = 80$
39. $20 + 2(8 - 5) + 1 = 20 + 2(3) + 1 = 20 + 6 + 1 = 27$
41. $5 + 2(3 \cdot 4 - 1) + 8 = 5 + 2(12 - 1) + 8 = 5 + 2(11) + 8 = 5 + 22 + 8 = 35$
43. $8 + 10 \div 2 = 8 + 5 = 13$
45. $4 + 8 \div 4 - 2 = 4 + 2 - 2 = 4$
47. $3 + 12 \div 3 + 6 \cdot 5 = 3 + 4 + 30 = 37$
49. $3 \cdot 8 + 10 \div 2 + 4 \cdot 2 = 24 + 5 + 8 = 37$
51. $(5 + 3)(5 - 3) = (8)(2) = 16$
53. $5^2 - 3^2 = 5 \cdot 5 - 3 \cdot 3 = 25 - 9 = 16$
55. $(4 + 5)^2 = 9^2 = 9 \cdot 9 = 81$
57. $4^2 + 5^2 = 4 \cdot 4 + 5 \cdot 5 = 16 + 25 = 41$
59. $3 \cdot 10^2 + 4 \cdot 10 + 5 = 300 + 40 + 5 = 345$
61. $2 \cdot 10^3 + 3 \cdot 10^2 + 4 \cdot 10 + 5 = 2000 + 300 + 40 + 5 = 2345$
63. $10 - 2(4 \cdot 5 - 16) = 10 - 2(20 - 16) = 10 - 2(4) = 10 - 8 = 2$
65. $4[7 + 3(2 \cdot 9 - 8)] = 4[7 + 3(18 - 8)] = 4[7 + 3(10)] = 4(7 + 30) = 4(37) = 148$
67. $5(7 - 3) + 8(6 - 4) = 5(4) + 8(2) = 20 + 16 = 36$
69. $3(4 \cdot 5 - 12) + 6(7 \cdot 6 - 40) = 3(20 - 12) + 6(42 - 40) = 3(8) + 6(2) = 24 + 12 = 36$
71. $3^4 + 4^2 \div 2^3 - 5^2 = 81 + 16 \div 8 - 25 = 81 + 2 - 25 = 58$
73. $5^2 + 3^4 \div 9^2 + 6^2 = 25 + 81 \div 81 + 36 = 25 + 1 + 36 = 62$
75. The next number is 5.
77. The next number is 10.
79. The next number is $5^2 = 25$.
81. Since $2 + 2 = 4$ and $2 + 4 = 6$ the next number is $4 + 6 = 10$.
83. (a) $6(100) = 600$ mg   (b) $2(45) + 3(47) = 231$ mg
85. See the table in the back of the textbook.
87. $2(10 + 4) = 2(14) = 28$ shares.
89. $3 \cdot 50 - 14 = 150 - 14 = 136$ dollars
91. $5 \cdot 2 = 10$ cookies in the package.
93. $210 \cdot 2 = 420$ calories.
95. $7 \cdot 32 = 224$ chips in the bag.
97. $80 + 15 = 95$ grams.

© 2000 Harcourt, Inc

1.2 Real Numbers

1. See the number line in the back of the textbook.
3. See the number line in the back of the textbook.
5. See the number line in the back of the textbook.
7. See the number line in the back of the textbook.

9. $\frac{3}{4} = \frac{3 \cdot 6}{4 \cdot 6} = \frac{18}{24}$

11. $\frac{1}{2} = \frac{1 \cdot 12}{2 \cdot 12} = \frac{12}{24}$

13. $\frac{5}{8} = \frac{5 \cdot 3}{8 \cdot 3} = \frac{15}{24}$

15. $\frac{3}{5} = \frac{3 \cdot 12}{5 \cdot 12} = \frac{36}{60}$

17. $\frac{11}{30} = \frac{11 \cdot 2}{30 \cdot 2} = \frac{22}{60}$

19. opposite -10, reciprocal $\frac{1}{10}$, absolute value 10.

21. opposite $-\frac{3}{4}$, reciprocal $\frac{4}{3}$, absolute value $\frac{3}{4}$.

23. opposite $-\frac{11}{2}$, reciprocal $\frac{2}{11}$, absolute value $\frac{11}{2}$.

25. opposite 3, reciprocal $-\frac{1}{3}$, absolute value 3.

27. opposite $\frac{2}{5}$, reciprocal $-\frac{5}{2}$, absolute value $\frac{2}{5}$.

29. opposite $-x$, reciprocal $\frac{1}{x}$, absolute value $|x|$ (The distance between x and 0 on the number line.)

31. $-5 < -3$

33. $-3 > -7$

35. Since $|-4| = 4$ and $-|-4| = -4$, $|-4| > -|-4|$

37. Since $-|-7| = -7$, $7 > -|-7|$

39. $-\frac{3}{4} < -\frac{1}{4}$

41. $-\frac{3}{2} < -\frac{3}{4}$

43. $|8 - 2| = |6| = 6$

45. $|5 \cdot 2^3 - 2 \cdot 3^2| = |5 \cdot 8 - 2 \cdot 9| = |40 - 18| = |22| = 22$

47. $|7 - 2| - |4 - 2| = |5| - |2| = 5 - 2 = 3$

49. $10 - |7 - 2(5 - 3)| = 10 - |7 - 2(2)| = 10 - |7 - 4| = 10 - |3| = 10 - 3 = 7$

51. $15 - |8 - 2(3 \cdot 4 - 9)| - 10 = 15 - |8 - 2(12 - 9)| - 10$
$= 15 - |8 - 2(3)| - 10$
$= 15 - |8 - 6| - 10$
$= 15 - |2| - 10$
$= 15 - 2 - 10$
$= 3$

53. $\frac{2}{3} \cdot \frac{4}{5} = \frac{2 \cdot 4}{3 \cdot 5} = \frac{8}{15}$

55. $\frac{1}{2}(3) = \frac{1}{2} \cdot \frac{3}{1} = \frac{3}{2}$

57. $\frac{1}{4}(5) = \frac{1}{4} \cdot \frac{5}{1} = \frac{1 \cdot 5}{4 \cdot 1} = \frac{5}{4}$

59. $\frac{4}{3} \cdot \frac{3}{4} = \frac{12}{12} = 1$

61. $6(\frac{1}{6}) = \frac{6}{1} \cdot \frac{1}{6} = \frac{6 \cdot 1}{1 \cdot 6} = \frac{6}{6} = 1$

63. $3 \cdot \frac{1}{3} = \frac{3}{1} \cdot \frac{1}{3} = \frac{3}{3} = 1$

65. $\left(\frac{3}{4}\right)^2 = \frac{3}{4} \cdot \frac{3}{4} = \frac{3 \cdot 3}{4 \cdot 4} = \frac{9}{16}$

67. $\left(\frac{2}{3}\right)^3 = \frac{2}{3} \cdot \frac{2}{3} \cdot \frac{2}{3} = \frac{8}{27}$

Problem Set 1.3

69. $\left(\frac{1}{10}\right)^4 = \frac{1}{10} \cdot \frac{1}{10} \cdot \frac{1}{10} \cdot \frac{1}{10} = \frac{1 \cdot 1 \cdot 1 \cdot 1}{10 \cdot 10 \cdot 10 \cdot 10} = \frac{1}{10,000}$

71. The next number is $\frac{1}{9}$.

73. The next number is $\frac{1}{5^2} = \frac{1}{25}$.

75. The perimeter is $4(1 \text{ in.}) = 4 \text{ in.}$, and the area is $(1 \text{ in.})^2 = 1 \text{ in.}^2$.

77. The perimeter is $2(1.5 \text{ inches}) + 2(0.75 \text{ inches}) = 3.0 \text{ inches} + 1.50 \text{ inches} = 4.5 \text{ inches}$
    and the area is $(1.5 \text{ inches})(0.75 \text{ inches}) = 1.125 \text{ inches}^2$.

79. The perimeter is $2.75 \text{ cm} + 4 \text{ cm} + 3.5 \text{ cm} = 10.25 \text{ cm}$, and the area is $\frac{1}{2}(4 \text{ cm})(2.5 \text{ cm}) = 5.0 \text{ cm}^2$.

81. A loss of 8 yds corresponds to −8 on the number line. The total yds gained corresponds to −2 yds.

83. The temperature can be represented as −64°. The new (warmer) temperature corresponds to −54°.

85. −15°

87. See the table in the back of the textbook.

89. His position corresponds to −100 feet. His new (deeper) position corresponds to −105 feet.

91. The area is given by: $8\frac{1}{2} \cdot 11 = \frac{17}{2} \cdot \frac{11}{1} = \frac{187}{2} = 93.5 \text{ in.}^2$.

93. The calories consumed would be: $2(544) + (299) = 1387$ calories

95. The calories consumed by the 180 lb person would be $3(653) = 1,959$ calories, while the calories consumed by the 120 lb person would be $3(435) = 1,305$ calories. Thus the 180 lb person consumed $1959 - 1305 = 654$ more calories.

1.3 Addition of Real Numbers

1. $3 + 5 = 8, \ 3 + (-5) = -2, \ -3 + 5 = 2, \ (-3) + (-5) = -8$

3. $15 + 20 = 35, \ 15 + (-20) = -5, \ -15 + 20 = -5, \ (-15) + (-20) = 35$

5. $6 + (-3) = 3$

7. $13 + (-20) = -7$

9. $18 + (-32) = -14$

11. $-6 + 3 = -3$

13. $-30 + 5 = -25$

15. $-6 + (-6) = -12$

17. $-9 + (-10) = -19$

19. $-10 + (-15) = -25$

21. $5 + (-6) + (-7) = 5 + (-13) = -8$

23. $-7 + 8 + (-5) = -12 + 8 = -4$

25. $5 + [6 + (-2)] + (-3) = 5 + 4 + (-3) = 9 + (-3) = 6$

27. $[6 + (-2)] + [3 + (-1)] = 4 + 2 = 6$

29. $20 + (-6) + [3 + (-9)] = 20 + (-6) + (-6) = 20 + (-12) = 8$

31. $-3 + (-2) + [5 + (-4)] = -3 + (-2) + 1 = -5 + 1 = -4$

33. $(-9 + 2) + [5 + (-8)] + (-4) = -7 + (-3) + (-4) + -14$

35. $[-6 + (-4)] + [7 + (-5)] + (-9) = -10 + 2 + (-9) = -19 + 2 = -17$

37. $(-6 + 9) + (-5) + (-4 + 3) + 7 = 3 + (-5) + (-1) + 7 = 10 + (-6) = 4$

39. $-5 + 2(-3 + 7) = -5 + 2(4) = -5 + 8 = 3$

41. $9 + 3(-8 + 10) = 9 + 3(2) = 9 + 6 = 15$

43. $-10 + 2(-6 + 8) + (-2) = -10 + 2(2) + (-2) = -10 + 4 + (-2) = -12 + 4 = -8$
45. $2(-4 + 7) + 3(-6 + 8) = 2(3) + 3(2) = 6 + 6 = 12$
47. The pattern is to add 5, so the next two terms are $18 + 5 = 23 + 5 = 28$.
49. The pattern is to add 5, so the next two terms are $25 + 5 = 30$ and $30 + 5 = 35$.
51. The pattern is to add -5, so the next two terms are $5 + (-5) = 0$ and $0 + (-5) = -5$.
53. The pattern is to add -6, so the next two terms are $-6 + (-6) = -12$ and $-12 + (-6) = -18$.
55. The pattern is to add -4, so the next two terms are $0 + (-4) = -4$ and $-4 + (-4) = -8$.
57. Yes, since each successive odd number is 2 added to the previous one.
59. $5 + 9 = 14$
61. $[-7 + (-5)] + 4 = -12 + 4 = -8$
63. $[-2 + (-3)] + 10 = -5 + 10 = 5$
65. The number is 3, since $-8 + 3 = -5$
67. The number is -3, since $-6 + (-3) = -9$
69. $-12° + 4° = -8°$.
71. $\$10 + (-\$6) + (-\$8) = \$10 + (-\$14) = -\$4$.
73. The new balance is $-\$30 + \$40 = \$10$.
75. The sequence of her wages is: $\$6.50, \$6.75, \$7.00, \$7.25, \$7.50, \$7.75, \$8.00$.
77. (a) See the table in the back of the textbook.
    (b) $36,036, + 42,862 = \$78,898$

1.4 Subtraction of Real Numbers

1. $5 - 8 = 5 + (-8) = -3$
3. $3 - 9 = 3 + (-9) = -6$
5. $5 - 5 = 5 + (-5) = 0$
7. $-8 - 2 = -8 + (-2) = -10$
9. $-4 - 12 = -4 + (-12) = -16$
11. $-6 - 6 = -6 + (-6) = -12$
13. $-8 - (-1) = -8 + 1 = -7$
15. $15 - (-20) = 15 + 20 = 35$
17. $-4 - (-4) = -4 + 4 = 0$
19. $3 - 2 - 5 = 3 + (-2) + (-5) = 3 + (-7) = -4$
21. $9 - 2 - 3 = 9 + (-2) + (-3) = 9 + (-5) = 4$
23. $-6 - 8 - 10 = -6 + (-8) + (-10) = -24$
25. $-22 + 4 - 10 = -22 + 4 + (-10) = -32 + 4 = -28$
27. $10 - (-20) - 5 = 10 + 20 + (-5) = 30 + (-5) = 25$
29. $8 - (2 - 3) - 5 = 8 - (-1) - 5 = 8 + 1 + (-5) = 9 + (-5) = 4$
31. $7 - (3 - 9) - 6 = 7 - (-6) - 6 = 7 + 6 + (-6) = 13 + (-6) = 7$
33. $5 - (-8 - 6) - 2 = 5 - (-14) - 2 = 5 + 14 + (-2) = 19 + (-2) = 17$
35. $-(5 - 7) - (2 - 8) = -(-2) - (-6) = 2 + 6 = 8$
37. $-(3 - 10) - (6 - 3) = -(-7) - 3 = 7 + (-3) = 4$
39. $16 - [(4 - 5) - 1] = 16 - (-1 - 1) = 16 - (-2) = 16 + 2 = 18$
41. $5 - [(2 - 3) - 4] = 5 - (-1 - 4) = 5 - (-5) = 5 + 5 = 10$
43. $21 - [-(3 - 4) - 2] - 5 = 21 - [-(-1) - 2] - 5 = 21 - (1 - 2) - 5 = 21 - (-1) - 5 = 21 + 1 + (-5) = 22 + (-5) = 17$
45. $2 \cdot 8 - 3 \cdot 5 = 16 - 15 = 16 + (-15) = 1$
47. $3 \cdot 5 - 2 \cdot 7 = 15 - 14 = 15 + (-14) = 1$
49. $5 \cdot 9 - 2 \cdot 3 - 6 \cdot 2 = 45 - 6 - 12 = 45 + (-6) + (-12) = 45 + (-18) = 27$
51. $3 \cdot 8 - 2 \cdot 4 - 6 \cdot 7 = 24 - 8 - 42 = 24 + (-8) + (-42) = 24 + (-50) = -26$
53. $2 \cdot 3^2 - 5 \cdot 2^2 = 2 \cdot 9 - 5 \cdot 4 = 18 - 20 = 18 + (-20) = -2$
55. $4 \cdot 3^3 - 5 \cdot 2^3 = 4 \cdot 27 - 5 \cdot 8 = 108 - 40 = 108 + (-40) = 68$
57. $-7 - 4 = -7 + (-4) = -11$
59. $12 - (-8) = 12 + 8 = 20$
61. $-5 - (-7) = -5 + 7 = 2$
63. $[4 + (-5)] - 17 = -1 - 17 = -1 + (-17) = -18$
65. $8 - 5 = 8 + (-5) = 3$
67. $-8 - 5 = -8 + (-5) = -13$
69. $8 - (-5) = 8 + 5 = 13$
71. The number is 10, since $8 - 10 = 8 + (-10) = -2$.
73. The number is -2, since $8 - (-2) = 8 + 2 = 10$.
75. $\$1,500 - \$730$

Problem Set 1.5

77. $-\$35 + \$15 - \$20 = -\$35 + (-\$20) + \$15 = -\$55 + \$15 = -\$40$
79. $\$98 - \$65 - \$53 = \$98 + (-\$65) + (-\$53) = \$98 + (-\$118) = -\$20$
81. The sequence of values is $4500, $3950, $3400, $2850, and $2300. This is an arithmetic sequence, since −$550 is added to each value to obtain the new value.
83. The difference is 1000 feet −231 feet = 769 feet.
85. He is 439 feet from the starting line. 87. 2 seconds have gone by.
89. (a) See the table in the back of the textbook.
    (b) $205 - 121 = 84$ million tons
91. (a) See the table in the back of the textbook.
    (b) $33 - 28 = 5$ cents per minute
93. The angles add to 90°, so $x = 90° - 55° = 35°$. 95. The angles add to 180°, so $x = 180° - 120° = 60°$.

1.5 Properties of Real Numbers

1. Commutative property of addition
3. Multiplicative inverse property
5. Commutative property of addition
7. Distributive property
9. Commutative and associative properties of addition
11. Commutative and associative properties of addition
13. Commutative property of addition
15. Commutative and associative properties of multiplication
17. Commutative property of multiplication
19. Additive inverse property
21. $3(x+2) = 3x + 6$
23. $9(a+b) = 9a + 9b$
25. $3(0) = 0$
27. $3 + (-3) = 0$
29. $10(1) = 10$
31. $4 + (2 + x) = (4 + 2) + x = 6 + x$
33. $(x + 2) + 7 = x + (2 + 7) = x + 9$
35. $3(5x) = (3 \cdot 5)x = 15x$
37. $9(6y) = (9 \cdot 6)y = 54y$
39. $\frac{1}{2}(3a) = (\frac{1}{2} \cdot 3)a = \frac{3}{2}a$
41. $\frac{1}{3}(3x) = (\frac{1}{3} \cdot 3)x = 1x = x$
43. $\frac{1}{2}(2y) = (\frac{1}{2} \cdot 2)y = 1y = y$
45. $\frac{3}{4}(\frac{4}{3}x) = (\frac{3}{4} \cdot \frac{4}{3})x = 1x = x$
47. $\frac{6}{5}(\frac{5}{6}a) = (\frac{6}{5} \cdot \frac{5}{6})a = 1a = a$
49. $8(x + 2) = 8 \cdot x + 8 \cdot 2 = 8x + 16$
51. $8(x - 2) = 8 \cdot x - 8 \cdot 2 = 8x - 16$
53. $4(y + 1) = 4 \cdot y + 4 \cdot 1 = 4y + 4$
55. $3(6x + 5) = 3 \cdot 6x + 3 \cdot 5 = 18x + 15$
57. $2(3a + 7) = 2 \cdot 3a + 2 \cdot 7 = 6a + 14$
59. $9(6y - 8) = 9 \cdot 6y - 9 \cdot 8 = 54y - 72$
61. $\frac{1}{2}(3x - 6) = \frac{1}{2}(3x) - \frac{1}{2}(6) = \frac{3}{2}x - 3$
63. $\frac{1}{3}(3x + 6) = \frac{1}{3} \cdot 3x + \frac{1}{3} \cdot 6 = x + 2$
65. $3(x + y) = 3x + 3y$
67. $8(a - b) = 8a - 8b$
69. $6(2x + 3y) = 6(2x) + 6(3y) = 12x + 18y$
71. $4(3a - 2b) = 4 \cdot 3a - 4 \cdot 2b = 12a - 8b$
73. $\frac{1}{2}(6x + 4y) = \frac{1}{2}(6x) + \frac{1}{2}(4y) = 3x + 2y$
75. $4(a + 4) + 9 = 4a + 16 + 9 = 4a + 25$
77. $2(3x + 5) + 2 = 6x + 10 + 2 = 6x + 12$
79. $7(2x + 4) + 10 = 14x + 28 + 10 = 14x + 38$
81. No. The man cannot reverse the order of putting on his socks and putting put on his shoes.
83. No. The skydiver must jump out of the plane before pulling the rip cord.
85. Division is not a commutative operation. For example, $8 \div 4 = 2$ while $4 \div 8 = \frac{1}{2}$.
87. $12(2400 - 480) = 12(1920) = \$23,040$
    $12 \cdot 2400 - 12 \cdot 480 = 28,800 - 5,760 = \$23,040$
89. $P = 2w + 2l = 2(w + l)$

© 2000 Harcourt, Inc

1.6 Multiplication of Real Numbers

1. $7(-6) = -42$
3. $-7(3) = -21$
5. $-8(2) = -16$
7. $-3(-1) = 3$
9. $-11(-11) = 121$
11. $-3(2)(-1) = 6$
13. $-3(-4)(-5) = -60$
15. $-2(-4)(-3)(-1) = 24$
17. $(-7)^2 = (-7)(-7) = 49$
19. $(-3)^3 = (-3)(-3)(-3) = -27$
21. $-2(2-5) = -2(-3) = 6$
23. $-5(8-10) = -5(-2) = 10$
25. $(4-7)(6-9) = (-3)(-3) = 9$
27. $(-3-2)(-5-4) = (-5)(-9) = 45$
29. $-3(-6) + 4(-1) = 18 + (-4) = 14$
31. $2(3) - 3(-4) + 4(-5) = 6 + 12 + (-20) = 18 + (-20) = -2$
33. $4(-3)^2 + 5(-6)^2 = 4(9) + 5(36) = 36 + 180 = 216$
35. $7(-2)^3 - 2(-3)^3 = 7(-8) - 2(-27) = -56 + 54 = -2$
37. $6 - 4(8-2) = 6 - 4(6) = 6 - 24 = 6 + (-24) = -18$
39. $9 - 4(3-8) = 9 - 4(-5) = 9 + 20 = 29$
41. $-4(3-8) - 6(2-5) = -4(-5) - 6(-3) = 20 + 18 = 38$
43. $7 - 2[-6 - 4(-3)] = 7 - 2(-6+12) = 7 - 2(6) = 7 - 12 = 7 + (-12) = -5$
45. $7 - 3[2(-4-4) - 3(-1-1)] = 7 - 3[2(-8) - 3(-2)] = 7 - 3(-16+6) = 7 - 3(-10) = 7 + 30 = 37$
47. $8 - 6[-2(-3-1) + 4(-2-3)] = 8 - 6[-2(-4) + 4(-5)] = 8 - 6[8 + (-20)] = 8 - 6(-12) = 8 + 72 = 80$
49. $-\frac{2}{3} \cdot \frac{5}{7} = -\frac{2 \cdot 5}{3 \cdot 7} = -\frac{10}{21}$
51. $-8(\frac{1}{2}) = -\frac{8}{1} \cdot \frac{1}{2} = -\frac{8}{2} = -4$
53. $-\frac{3}{4}(-\frac{4}{3}) = -\frac{3}{4} \cdot (-\frac{4}{3}) = \frac{12}{12} = 1$
55. $(-\frac{3}{4})^2 = (-\frac{3}{4})(-\frac{3}{4}) = \frac{9}{16}$
57. $(-\frac{2}{3})^3 = (-\frac{2}{3})(-\frac{2}{3})(-\frac{2}{3}) = -\frac{8}{27}$
59. $-2(4x) = (-2 \cdot 4)x = -8x$
61. $-7(-6x) = [-7(-6)]x = 42x$
63. $-\frac{1}{3}(-3x) = [-\frac{1}{3} \cdot (-3)]x = 1x = x$
65. $-4(-\frac{1}{4}x) = [-4(-\frac{1}{4})]x = 1x = x$
67. $-4(a+2) = -4a + (-4)(2) = -4a - 8$
69. $-\frac{1}{2}(3x-6) = -\frac{3}{2}x - \frac{1}{2}(-6) = -\frac{3}{2}x + 3$
71. $-3(2x-5) - 7 = -6x + 15 - 7 = -6x + 8$
73. $-5(3x+4) - 10 = -15x - 20 - 10 = -15x - 30$
75. $3(-10) + 5 = -30 + 5 = -25$
77. $2(-4x) = -8x$
79. $-9 \cdot 2 - 8 = -18 + (-8) = -26$

81. The pattern is to multiply by 2, so the next number is $4 \cdot 2 = 8$.
83. The pattern is to multiply by -2, so the next number is $40 \cdot (-2) = -80$.
85. The pattern is to multiply by $\frac{1}{2}$, so the next number is $\frac{1}{4} \cdot \frac{1}{2} = \frac{1}{8}$.
87. The pattern is to multiply by $\frac{1}{2}$, so the next number is $2 \cdot \frac{1}{2} = 1$.
89. The pattern is to multiply by -2, so the next number is $12 \cdot (-2) = -24$.
91. The amount lost is: $20(\$3) = \$60$.
93. The temperature is: $25° - 4(6°) = 25° - 24° = 1°$
95. $500, $1000, $2000, $4000, $8000, and $16,000. Yes, this is a geometric sequence, since each value is 2 times the preceeding value.
97. $2(630) - 3(265) = +465$ calories (gain)

Problem Set 1.7

1.7 Division of Real Numbers

1. $\frac{8}{-4} = -2$
3. $\frac{-48}{16} = -3$
5. $\frac{-7}{21} = -\frac{1}{3}$
7. $\frac{-39}{-13} = 3$
9. $\frac{-6}{-42} = \frac{1}{7}$
11. $\frac{0}{-32} = 0$
13. $-3 + 12 = 9$
15. $-3 - 12 = -3 + (-12) = -15$
17. $-3(12) = -36$
19. $-3 \div 12 = \frac{-3}{12} = -\frac{1}{4}$
21. $\frac{4}{5} \div \frac{3}{4} = \frac{4}{5} \cdot \frac{4}{3} = \frac{16}{15}$
23. $-\frac{5}{6} \div (-\frac{5}{8}) = -\frac{5}{6} \cdot (-\frac{8}{5}) = \frac{40}{30} = \frac{4}{3}$
25. $\frac{10}{13} \div (-\frac{5}{4}) = \frac{10}{13} \cdot (-\frac{4}{5}) = -\frac{40}{65} = -\frac{8}{13}$
27. $-\frac{5}{6} \div \frac{5}{6} = -\frac{5}{6} \cdot \frac{6}{5} = -\frac{30}{30} = -1$
29. $-\frac{3}{4} \div (-\frac{3}{4}) = -\frac{3}{4} \cdot (-\frac{4}{3}) = \frac{12}{12} = 1$
31. $\frac{3(-2)}{-10} = \frac{-6}{-10} = \frac{3}{5}$
33. $\frac{-5(-5)}{-15} = \frac{25}{-15} = -\frac{5}{3}$
35. $\frac{-8(-7)}{-28} = \frac{56}{-28} = -2$
37. $\frac{27}{4-13} = \frac{27}{-9} = -3$
39. $\frac{20-6}{5-5} = \frac{14}{0} =$ undefined
41. $\frac{-3+9}{2(5)-10} = \frac{6}{10-10} = \frac{6}{0} =$ undefined
43. $\frac{15(-5)-25}{2(-10)} = \frac{-75-25}{-20} = \frac{-100}{-20} = 5$
45. $\frac{27-2(-4)}{-3(5)} = \frac{27+8}{-15} = \frac{35}{-15} = -\frac{7}{3}$
47. $\frac{12-6(-2)}{12(-2)} = \frac{12+12}{-24} = \frac{24}{-24} = -1$
49. $\frac{5^2-2^2}{-5+2} = \frac{25-4}{-3} = \frac{21}{-3} = -7$
51. $\frac{8^2-2^2}{8^2+2^2} = \frac{64-4}{64+4} = \frac{60}{68} = \frac{15}{17}$
53. $\frac{(5+3)^2}{-5^2-3^2} = \frac{8^2}{-25-9} = \frac{64}{-34} = -\frac{32}{17}$
55. $\frac{(8-4)^2}{8^2-4^2} = \frac{4^2}{64-16} = \frac{16}{48} = \frac{1}{3}$
57. $\frac{-4 \cdot 3^2 - 5 \cdot 2^2}{-8(7)} = \frac{-4 \cdot 9 - 5 \cdot 4}{-56} = \frac{-36-20}{-56} = \frac{-56}{-56} = 1$
59. $\frac{3 \cdot 10^2 + 4 \cdot 10 + 5}{345} = \frac{300+40+5}{345} = \frac{345}{345} = 1$
61. $\frac{7-[(2-3)-4]}{-1-2-3} = \frac{7-(-1-4)}{-6} = \frac{7-(-5)}{-6} = \frac{7+5}{-6} = \frac{12}{-6} = -2$
63. $\frac{6(-4)-2(5-8)}{-6-3-5} = \frac{-24-2(-3)}{-14} = \frac{-24+6}{-14} = \frac{-18}{-14} = \frac{9}{7}$
65. $\frac{3(-5-3)+4(7-9)}{5(-2)+3(-4)} = \frac{3(-8)+4(-2)}{-10+(-12)} = \frac{-24+(-8)}{-22} = \frac{-32}{-22} = \frac{16}{11}$
67. $\frac{|3-9|}{3-9} = \frac{|-6|}{-6} = \frac{6}{-6} = -1$
69. The quotient is $\frac{-12}{-4} = 3$.
71. The number is -10, since $\frac{-10}{-5} = 2$.
73. The number is -3, since $\frac{27}{-3} = -9$.
75. The expression is: $\frac{-20}{4} - 3 = -5 - 3 = -8$
77. Each person would lose: $\frac{13600-15000}{4} = \frac{-1400}{4} = -350 = \$350$ loss per person

© 2000 Harcourt, Inc

79. The change per hour is: $\frac{61°-75°}{4} = \frac{-14°}{4} = -3.5$ Drops 3.5° F. each hour.

1.8 Subsets of the Real Numbers

1. The whole numbers are: 0, 1
3. The rational numbers are: $-3, -2.5, 0, 1, \frac{3}{2}$
5. The real numbers are: $\{-3, -2.5, 0, 1, \frac{3}{2}, \sqrt{15}\}$
7. The integers are: $-10, -8, -2, 9$
9. The irrational numbers are: π
11. True
13. False
15. False
17. True
19. This number is composite: $48 = 6 \cdot 8 = (2 \cdot 3) \cdot (2 \cdot 2 \cdot 2) = 2^4 \cdot 3$
21. This number is prime.
23. This number is composite: $1023 = 3 \cdot 341 = 3 \cdot 11 \cdot 31$
25. $144 = 12 \cdot 12 = (3 \cdot 4) \cdot (3 \cdot 4) = (3 \cdot 2 \cdot 2) \cdot (3 \cdot 2 \cdot 2) = 2^4 \cdot 3^2$
27. $38 = 2 \cdot 19$
29. $105 = 5 \cdot 21 = 5 \cdot (3 \cdot 7) = 3 \cdot 5 \cdot 7$
31. $180 = 10 \cdot 18 = (2 \cdot 5) \cdot (3 \cdot 2 \cdot 3) = 2^2 \cdot 3^2 \cdot 5$
33. $385 = 5 \cdot 77 = 5 \cdot (7 \cdot 11) = 5 \cdot 7 \cdot 11$
35. $121 = 11 \cdot 11 = 11^2$
37. $420 = 10 \cdot 42 = (2 \cdot 5) \cdot (7 \cdot 6) = (2 \cdot 5) \cdot (7 \cdot 2 \cdot 3) = 2^2 \cdot 3 \cdot 5 \cdot 7$
39. $620 = 10 \cdot 62 = (2 \cdot 5) \cdot (2 \cdot 31) = 2^2 \cdot 5 \cdot 31$
41. $\frac{105}{165} = \frac{3 \cdot 5 \cdot 7}{3 \cdot 5 \cdot 11} = \frac{7}{11}$
43. $\frac{525}{735} = \frac{3 \cdot 5 \cdot 5 \cdot 7}{3 \cdot 5 \cdot 7 \cdot 7} = \frac{5}{7}$
45. $\frac{385}{455} = \frac{5 \cdot 7 \cdot 11}{5 \cdot 7 \cdot 13} = \frac{11}{13}$
47. $\frac{322}{345} = \frac{2 \cdot 7 \cdot 23}{3 \cdot 5 \cdot 23} = \frac{2 \cdot 7}{3 \cdot 5} = \frac{14}{15}$
49. $\frac{205}{369} = \frac{5 \cdot 41}{3 \cdot 3 \cdot 41} = \frac{5}{3 \cdot 3} = \frac{5}{9}$
51. $\frac{215}{344} = \frac{5 \cdot 43}{2 \cdot 2 \cdot 2 \cdot 43} = \frac{5}{2 \cdot 2 \cdot 2} = \frac{5}{8}$
53. $6^3 = (2 \cdot 3)^3 = 2^3 \cdot 3^3$
55. $9^4 \cdot 16^2 = (3 \cdot 3)^4 \cdot (2 \cdot 2 \cdot 2 \cdot 2)^2 = 2^8 \cdot 3^8$
57. $3 \cdot 8 + 3 \cdot 7 + 3 \cdot 5 = 24 + 21 + 15 = 60 = 6 \cdot 10 = (2 \cdot 3) \cdot (2 \cdot 5) = 2^2 \cdot 3 \cdot 5$
59. They are not a subset of the irrational numbers.
61. 8, 21, 34

Problem Set 1.9

1.9 Addition and Subtraction with Fractions

1. $\frac{3}{6} + \frac{1}{6} = \frac{4}{6} = \frac{2}{3}$

3. $\frac{3}{8} - \frac{5}{8} = -\frac{2}{8} = -\frac{1}{4}$

5. $-\frac{1}{4} + \frac{3}{4} = \frac{2}{4} = \frac{1}{2}$

7. $\frac{x}{3} - \frac{1}{3} = \frac{x-1}{3}$

9. $\frac{1}{4} + \frac{2}{4} + \frac{3}{4} = \frac{6}{4} = \frac{3}{2}$

11. $\frac{x+7}{2} - \frac{1}{2} = \frac{x+7-1}{2} = \frac{x+6}{2}$

13. $\frac{1}{10} - \frac{3}{10} - \frac{4}{10} = -\frac{6}{10} = -\frac{3}{5}$

15. $\frac{1}{a} + \frac{4}{a} + \frac{5}{a} = \frac{10}{a}$

17. $\frac{1}{8} + \frac{3}{4} = \frac{1}{8} + \frac{3 \cdot 2}{4 \cdot 2} = \frac{1}{8} + \frac{6}{8} = \frac{7}{8}$

19. $\frac{3}{10} - \frac{1}{5} = \frac{3}{10} - \frac{1 \cdot 2}{5 \cdot 2} = \frac{3}{10} - \frac{2}{10} = \frac{1}{10}$

21. $\frac{4}{9} + \frac{1}{3} = \frac{4}{9} + \frac{1 \cdot 3}{3 \cdot 3} = \frac{4}{9} + \frac{3}{9} = \frac{7}{9}$

23. $2 + \frac{1}{3} = \frac{2 \cdot 3}{1 \cdot 3} + \frac{1}{3} = \frac{6}{3} + \frac{1}{3} = \frac{7}{3}$

25. $-\frac{3}{4} + 1 = -\frac{3}{4} + \frac{1 \cdot 4}{1 \cdot 4} = -\frac{3}{4} + \frac{4}{4} = \frac{1}{4}$

27. $\frac{1}{2} + \frac{2}{3} = \frac{1 \cdot 3}{2 \cdot 3} + \frac{2 \cdot 2}{3 \cdot 2} = \frac{3}{6} + \frac{4}{6} = \frac{7}{6}$

29. $\frac{5}{12} - (-\frac{3}{8}) = \frac{5}{12} + \frac{3}{8} = \frac{5 \cdot 2}{12 \cdot 2} + \frac{3 \cdot 3}{8 \cdot 3} = \frac{10}{24} + \frac{9}{24} = \frac{19}{24}$

31. $-\frac{1}{20} + \frac{8}{30} = -\frac{1 \cdot 3}{20 \cdot 3} + \frac{8 \cdot 2}{30 \cdot 2} = -\frac{3}{60} + \frac{16}{60} = \frac{13}{60}$

33. First factor the denominators to find the LCM:

   $30 = 2 \cdot 3 \cdot 5$

   $42 = 2 \cdot 3 \cdot 7$

   LCM $= 2 \cdot 3 \cdot 5 \cdot 7 = 210$

   Combining the fractions: $\frac{17}{30} + \frac{11}{42} = \frac{17 \cdot 7}{30 \cdot 7} + \frac{11 \cdot 5}{42 \cdot 5} = \frac{119}{210} + \frac{55}{210} = \frac{174}{210} = \frac{2 \cdot 3 \cdot 29}{2 \cdot 3 \cdot 5 \cdot 7} = \frac{29}{5 \cdot 7} = \frac{29}{35}$

35. First factor the denominators to find the LCM:

   $84 = 2 \cdot 2 \cdot 3 \cdot 7$

   $90 = 2 \cdot 3 \cdot 3 \cdot 5$

   LCM $= 2 \cdot 2 \cdot 3 \cdot 3 \cdot 5 \cdot 7 = 1260$

   Combining the fractions: $\frac{25}{84} + \frac{41}{90} = \frac{25 \cdot 15}{84 \cdot 15} + \frac{41 \cdot 14}{90 \cdot 14} = \frac{375}{1260} + \frac{574}{1260} = \frac{949}{1260}$

37. First factor the denominators to find the LCM:

   $126 = 2 \cdot 3 \cdot 3 \cdot 7$

   $180 = 2 \cdot 2 \cdot 3 \cdot 3 \cdot 5$

   LCM $= 2 \cdot 2 \cdot 3 \cdot 3 \cdot 5 \cdot 7 = 1260$

   Combining the fractions:

   $\frac{13}{126} - \frac{13}{180} = \frac{13 \cdot 10}{126 \cdot 10} - \frac{13 \cdot 7}{180 \cdot 7} = \frac{130}{1260} - \frac{91}{1260} = \frac{39}{1260} = \frac{3 \cdot 13}{2 \cdot 2 \cdot 3 \cdot 3 \cdot 5 \cdot 7} = \frac{13}{2 \cdot 2 \cdot 3 \cdot 5 \cdot 7} = \frac{13}{420}$

© 2000 Harcourt, Inc

39. Combining the fractions: $\frac{3}{4} + \frac{1}{8} + \frac{5}{6} = \frac{3 \cdot 6}{4 \cdot 6} + \frac{1 \cdot 3}{8 \cdot 3} + \frac{5 \cdot 4}{6 \cdot 4} = \frac{18}{24} + \frac{3}{24} + \frac{20}{24} = \frac{41}{24}$

41. Combining the fractions: $\frac{1}{2} + \frac{1}{3} + \frac{1}{4} + \frac{1}{6} = \frac{1 \cdot 6}{2 \cdot 6} + \frac{1 \cdot 4}{3 \cdot 4} + \frac{1 \cdot 3}{4 \cdot 3} + \frac{1 \cdot 2}{6 \cdot 2} = \frac{6}{12} + \frac{4}{12} + \frac{3}{12} + \frac{2}{12} = \frac{15}{12} = \frac{5}{4}$

43. The sum is given by: $\frac{3}{7} + 2 + \frac{1}{9} = \frac{3 \cdot 9}{7 \cdot 9} + \frac{2 \cdot 63}{1 \cdot 63} + \frac{1 \cdot 7}{9 \cdot 7} = \frac{27}{63} + \frac{126}{63} + \frac{7}{63} = \frac{160}{63}$

45. The difference is given by: $\frac{7}{8} - \frac{1}{4} = \frac{7}{8} - \frac{1 \cdot 2}{4 \cdot 2} = \frac{7}{8} - \frac{2}{8} = \frac{5}{8}$

47. The pattern is to add $-\frac{1}{3}$, so the fourth term is: $-\frac{1}{3} + \left(-\frac{1}{3}\right) = -\frac{2}{3}$

49. The pattern is to add $\frac{2}{3}$, so the fourth term is: $\frac{5}{3} + \frac{2}{3} = \frac{7}{3}$

51. The pattern is to multiply by $\frac{1}{5}$, so the fourth term is: $\frac{1}{25} \cdot \frac{1}{5} = \frac{1}{125}$

CHAPTER 1 REVIEW

1. $-7 + (-10) = -17$
3. $(-3 + 12) + 5 = 9 + 5 = 14$
5. $9 - (-3) = 9 + 3 = 12$
7. $(-3)(-7) - 6 = 21 - 6 = 15$
9. $2(-8 \cdot 3x) = 2(-24x) = -48x$
11. $(-40/8) - 7 = -5 - 7 = -5 + (-7) = -12$

13-17 See the graph in the back of the textbook.

19. $|12| = 12$
21. $\left|-\frac{4}{5}\right| = \frac{4}{5}$
23. $|-1.8| = 1.8$
25. The opposite is -6, and the reciprocal is $\frac{1}{6}$.
27. The opposite is 9, and the reciprocal is $-\frac{1}{9}$.
29. $\left(\frac{2}{5}\right)\left(\frac{3}{7}\right) = \frac{6}{35}$
31. $\left(-\frac{4}{5}\right)\left(\frac{25}{16}\right) = -\frac{4 \cdot 25}{5 \cdot 16} = -\frac{100}{80} = -\frac{5}{4}$
33. $-18 + (-20) = -38$
35. $(-5) + (-10) + (-7) = -15 + (-7) = -22$
37. $(-21) + 40 + (-23) + 5 = 19 + (-23) + 5 = -4 + 5 = 1$
39. $14 - (-8) = 14 + 8 = 22$
41. $4 - 9 - 15 = 4 + (-9) + (-15) = -5 + (-15) = -20$
43. $5 - (-10 - 2) - 3 = 5 - [-10 + (-2)] - 3 = 5 - (-12) - 3 = 5 + 12 + (-3) = 17 + (-3) = 14$
45. $20 - [-(10 - 3) - 8] - 7 = 20 - [-(7) - 8] - 7 = 20 - [-7 + (-8)] - 7 = 20 - [-15] - 7 = 35 + (-7) = 28$
47. $4(-3) = -12$
49. $(-1)(-3)(-1)(-4) = 3(-1)(-4) = (-3)(-4) = 12$
51. $\frac{-9}{36} = -\frac{1}{4}$
53. $4 \cdot 5 + 3 = 20 + 3 = 23$
55. $2^3 - 4 \cdot 3^2 + 5^2 = 8 - 4 \cdot 9 + 25 = 8 - 36 + 25 = 8 + (-36) + 25 = -28 + 25 = -3$
57. $20 + 8 \div 4 + 2 \cdot 5 = 20 + 2 + 10 = 22 + 10 = 32$
59. $-4(-5) + 10 = 20 + 10 = 30$

61. $3(4 - 7)^2 - 5(3 - 8)^2 = 3[4 + (-7)]^2 - 5[3 + (-8)]^2$
$= 3(-3)^2 - 5(-5)^2$
$= 3(9) - 5(25)$
$= 27 - 125$
$= 27 + (-125)$
$= -98$

© 2000 Harcourt, Inc

Chapter 1 Review

63. $\frac{4(-3)}{-6} = \frac{-12}{-6} = \frac{12}{6} = 2$

65. $\frac{15-10}{6-6} = \frac{5}{0} =$ undefined

67. $\frac{2(-7)+(-11)(-4)}{7-(-3)} = \frac{-14+44}{10} = \frac{30}{10} = 3$

69. $8(1) = 8$ multiplicative identity

71. $5 + (-5) = 0$ additive inverse

73. $8 + 0 = 8$ additive identity

75. $5(w - 6) = 5w - 30$ distributive property

77. $4(7a) = (4 \cdot 7)a = 28a$

79. $\frac{4}{5}(\frac{5}{4}y) = (\frac{4}{5} \cdot \frac{5}{4})y = \frac{20}{20}y = 1y = y$

81. $3(2a - 4) = 3(2a) - 3(4) = 6a - 12$

83. $(-1/2)(3x - 6) = (-1/2)(3x) - (-1/2)(6)$
$= (-1/2)(3x/1) - (-1/2)(6/1)$
$= (-3x/2) - (-6/2)$
$= (-3/2x) + 3$

85. 0, 5 Remember: Counting numbers and the number 0 are integers.

87. 0, 5, -3   Remember: Whole numbers and the opposites of all the counting numbers

89. $840 = 84 \cdot 10$
$= 7 \cdot 12 \cdot 2 \cdot 5$
$= 7 \cdot 3 \cdot 4 \cdot 2 \cdot 5$
$= 7 \cdot 3 \cdot 2 \cdot 2 \cdot 2 \cdot 5$
$= 2^3 \cdot 3 \cdot 5 \cdot 7$

91. $\frac{9}{70} + \frac{11}{84} = \frac{9 \cdot 6}{70 \cdot 6} + \frac{11 \cdot 5}{84 \cdot 5}$
$= \frac{54}{420} + \frac{55}{420}$
$= \frac{109}{420}$

$70 = 2 \cdot 5 \cdot 7$

$84 = 2^2 \cdot 3 \cdot 7$

LCD $= 2^2 \cdot 3 \cdot 5 \cdot 7 = 420$

93. 10, -30, 90, -270, 810, . . . For each new number multiply the previous number by −3.

95. 4, 6, 8, 10, 12, . . . For each new number add 2 to the previous number.

97. 1, -1/2, 1/4, -1/8, 1/16, . . . For each new number multiply the previous number by −1/2.

© 2000 Harcourt, Inc

CHAPTER 1 TEST

1. $x + 3 = 8$

2. $5y = 15$

3. $5^2 + 3(9-7) + 3^2 = 5^2 + 3(2) + 3^2 = 25 + 6 + 9 = 40$

4. $10 - 6 \div 3 + 2^3 = 10 - 6 \div 3 + 8 = 10 - 2 + 8 = 18 - 2 = 16$

5. Opposite 4, Reciprocal $-\frac{1}{4}$, Absolute value $|-4| = 4$

6. Opposite $-\frac{3}{4}$, Reciprocal $\frac{4}{3}$, Absolute value $\left|\frac{3}{4}\right| = \frac{3}{4}$

7. $3 + (-7) = -4$

8. $|-9 + (-6)| + |-3 + 5| = |-15| + |2| = 15 + 2 = 17$

9. $-4 - 8 = -4 + (-8) = -12$

10. $9 - (7-2) - 4 = 9 - 5 - 4 = 9 + (-5) + (-4) = 9 + (-9) = 0$

11. c. (associative property of addition)

12. e. (distributive property)

13. d. (associative property of multiplication)

14. a. (commutative property of addition)

15. $-3(7) = -21$

16. $-4(8)(-2) = 64$

17. $8\left(-\frac{1}{4}\right) = \frac{8}{1}\left(-\frac{1}{4}\right) = -\frac{8}{4} = -2$

18. $\left(-\frac{2}{3}\right)^3 = \left(-\frac{2}{3}\right)\left(-\frac{2}{3}\right)\left(-\frac{2}{3}\right) = -\frac{8}{27}$

19. $-3(-4) - 8 = 12 - 8 = 4$

20. $5(-6)^2 - 3(-2)^3 = 5(36) - 3(-8) = 180 + 24 = 204$

21. $7 - 3(2 - 8) = 7 - 3(-6) = 7 + 18 = 25$

22. $4 - 2[-3(-1+5) + 4(-3)] = 4 - 2[-3(4) + 4(-3)] = 4 - 2(-12 - 12) = 4 - 2(-24) = 4 + 48 = 52$

23. $\frac{4(-5) - 2(7)}{-10 - 7} = \frac{-20 - 14}{-17} = \frac{-34}{-17} = 2$

24. $\frac{2(-3-1) + 4(-5+2)}{-3(2) - 4} = \frac{2(-4) + 4(-3)}{-6 - 4} = \frac{-8 - 12}{-10} = \frac{-20}{-10} = 2$

25. $3 + (5 + 2x) = (3 + 5) + 2x = 8 + 2x$

26. $-2(-5x) = [-2(-5)]x = 10x$

27. $2(3x + 5) = 2(3x) + 2(5) = 6x + 10$

28. $(-1/2)(4x - 2) = (-1/2)(4x) + (-1/2) - 2 = -2x + 1$

29. The integers are 1 and -8.

30. The rational numbers are 1, 1.5, 3/4, -8.

31. The irrational numbers are $\sqrt{2}$.

32. The real numbers are 1, 1.5, $\sqrt{2}$, 3/4, and -8.

33. $592 = 4 \cdot 148 = (2 \cdot 2) \cdot (4 \cdot 37) = (2 \cdot 2) \cdot (2 \cdot 2 \cdot 37) = 2^4 \cdot 37$

34. $1340 = 10 \cdot 134 = (2 \cdot 5) \cdot (2 \cdot 67) = 2^2 \cdot 5 \cdot 67$

Chapter 1 Test

35. First factor the denominators to find the LCM:

    $15 = 3 \cdot 5$

    $42 = 2 \cdot 3 \cdot 7$

    LCM $= 2 \cdot 3 \cdot 5 \cdot 7 = 210$

    Combining the fractions: $\frac{5}{15} + \frac{11}{42} = \frac{5 \cdot 14}{15 \cdot 14} + \frac{11 \cdot 5}{42 \cdot 5} = \frac{70}{210} + \frac{55}{210} = \frac{125}{210} = \frac{5 \cdot 25}{5 \cdot 42} = \frac{25}{42}$

36. Combining the fractions: $\frac{5}{x} + \frac{3}{2} = \frac{5 \cdot 2}{x \cdot 2} + \frac{3 \cdot x}{2 \cdot x} = \frac{10}{2x} + \frac{3x}{2x} = \frac{10 + 3x}{2x}$

37. $8 + (-3) = 5$

38. $-24 - 2 = -24 + (-2) = -26$

39. $-5(-4) = 20$

40. $\frac{-24}{-2} = 12$

41. The pattern is to add 5, so the next term is $7 + 5 = 12$.

42. The pattern is to multiply by $-\frac{1}{2}$ so the next term is $-1(-\frac{1}{2}) = \frac{1}{2}$.

© 2000 Harcourt, Inc

Chapter 2
Linear Equations and Inequalities
2.1 Simplifying Expressions

1. $3x - 6x = (3-6)x = -3x$

3. $-2a + a = (-2+1)a = -a$

5. $7x + 3x + 2x = (7+3+2)x = 12x$

7. $3a - 2a + 5a = (3-2+5)a = 6a$

9. $4x - 3 + 2x = 4x + 2x - 3 = 6x - 3$

11. $3a + 4a + 5 = 7a + 5$

13. $2x - 3 + 3x - 2 = 2x + 3x - 3 - 2 = 5x - 5$

15. $3a - 1 + a + 3 = 3a + a - 1 + 3 = 4a + 2$

17. $-4x + 8 - 5x - 10 = -4x - 5x + 8 - 10 = -9x - 2$

19. $7a + 3 + 2a + 3a = 7a + 2a + 3a + 3 = 12a + 3$

21. $5(2x-1) + 4 = 10x - 5 + 4 = 10x - 1$

23. $7(3y+2) - 8 = 21y + 14 - 8 = 21y + 6$

25. $-3(2x-1) + 5 = -6x + 3 + 5 = -6x + 8$

27. $5 - 2(a+1) = 5 - 2a - 2 = -2a - 2 + 5 = -2a + 3$

29. $6 - 4(x-5) = 6 - 4x + 20 = -4x + 20 + 6 = -4x + 26$

31. $-9 - 4(2-y) + 1 = -9 - 8 + 4y + 1 = 4y + 1 - 9 - 8 = 4y - 16$

33. $-6 + 2(2-3x) + 1 = -6 + 4 - 6x + 1 = -6x - 6 + 4 + 1 = -6x - 1$

35. $(4x-7) - (2x+5) = 4x - 7 - 2x - 5 = 4x - 2x - 7 - 5 = 2x - 12$

37. $8(2a+4) - (6a-1) = 16a + 32 - 6a + 1 = 16a - 6a + 32 + 1 = 10a + 33$

39. $3(x-2) + (x-3) = 3x - 6 + x - 3 = 3x + x - 6 - 3 = 4x - 9$

41. $4(2y-8) - (y+7) = 8y - 32 - y - 7 = 8y - y - 32 - 7 = 7y - 39$

43. $-9(2x+1) - (x+5) = -18x - 9 - x - 5 = -18x - x - 9 - 5 = -19x - 14$

45. When $x = 2$: $3x - 1 = 3(2) - 1 = 6 - 1 = 5$

47. When $x = 2$: $-2x - 5 = -2(2) - 5 = -4 - 5 = -9$

49. When $x = 2$: $x^2 - 8x + 16 = (2)^2 - 8(2) + 16 = 4 - 16 + 16 = 4$

51. When $x = 2$: $(x-4)^2 = (2-4)^2 = (-2)^2 = 4$

53. When $x = -5$: $7x - 4 - x - 3 = 7(-5) - 4 - (-5) - 3 = -35 - 4 + 5 - 3 = -42 + 5 = -37$

    Now simplifying the expression: $7x - 4 - x - 3 = 7x - x - 4 - 3 = 6x - 7$

    When $x = -5$: $6x - 7 = 6(-5) - 7 = -30 - 7 = -37$

    Note that the two values are the same.

© 2000 Harcourt, Inc

Problem Set 2.1

55. When $x = -5$: $5(2x+1)+4 = 5[2(-5)+1]+4 = 5(-10+1)+4 = 5(-9)+4 = -45+4 = -41$

   Now simplifying the expression: $5(2x+1)+4 = 10x+5+4 = 10x+9$

   When $x = -5$: $10x+9 = 10(-5)+9 = -50+9 = -41$

57. When $x = -3$ and $y = 5$, $x^2 - 2xy + y^2 = (-3)^2 - 2(-3)(5) + (5)^2 = 9 + 30 + 25 = 64$

59. When $x = -3$ and $y = 5$: $(x-y)^2 = (-3-5)^2 = (-8)^2 = 64$

61. When $x = -3$ and $y = 5$, $x^2 + 6xy + 9y^2 = (-3)^2 + 6(-3)(5) + 9(5)^2 = 9 - 90 + 225 = 144$

63. When $x = -3$ and $y = 5$: $(x+3y)^2 = [-3+3(5)]^2 = (-3+15)^2 = (12)^2 = 144$

65. When $x = \frac{1}{2}$, $12x - 3 = 12(\frac{1}{2}) - 3 = 6 - 3 = 3$

67. When $x = \frac{1}{4}$: $12x - 3 = 12(\frac{1}{4}) - 3 = 3 - 3 = 0$

69. When $x = \frac{3}{2}$, $12x - 3 = 12(\frac{3}{2}) - 3 = 18 - 3 = 15$

71. When $x = \frac{3}{4}$: $12x - 3 = 12(\frac{3}{4}) - 3 = 9 - 3 = 6$

73. $2n + 3$ when $n = 1$, $2(1) + 3 = 5$

   $2n + 3$ when $n = 2$, $2(2) + 3 = 7$

   $2n + 3$ when $n = 3$, $2(3) + 3 = 9$

   $2n + 3$ when $n = 4$, $2(4) + 3 = 11$

75. $n^2 + 1$ when $n = 1$, $(1)^2 + 1 = 2$

   $n^2 + 1$ when $n = 2$, $(2)^2 + 1 = 5$

   $n^2 + 1$ when $n = 3$, $(3)^2 + 1 = 10$

   $n^2 + 1$ when $n = 4$, $(4)^2 + 1 = 17$

© 2000 Harcourt, Inc

77. See the tables in the back of the textbook.

79. $T = -0.0035A + 70$
   (a) When $A = 8,000$ feet
      $T = -0.0035(8,000) + 70$
      $= 42°\text{ F}$
   (b) When $A = 12,000$ feet
      $T = -00.0035(12,000) + 70$
      $= 28°\text{ F}$
   (c) when $A = 24,000$ feet
      $T = -0.0035(24,000) + 70$
      $= -14°\text{ F}$

81. $C = 35 + 0.25t$
   (a) When $t = 10$
      $C = 35 + 0.25(10)$
      $= \$37.50$
   (b) When $t = 20$
      $C = 35 + 0.25(20)$
      $= \$40.00$
   (c) When $t = 30$
      $C = 35 + 0.25(30)$
      $= \$42.50$

83. $T = G - 0.21G - 0.08G$
     $= 0.71G$
  When $G = 1,250$
     $T = 0.71(1,250)$
       $= \$887.50$

85. $x - 5$
  When $x = -2$: $x - 5 = (-2) - 5 = -7$

87. $2(x + 10)$
  When $x = -2$: $2(x + 10) = 2(-2 + 10)$
                      $= 2(8) = 16$

89. $\dfrac{10}{x}$
  When $x = -2$: $\dfrac{10}{x} = \dfrac{10}{-2} = -5$

91. $[3x + (-2)] - 5 = 3x - 2 - 5 = 3x - 7$
  When $x = -2$: $3x - 7 = 3(-2) - 7$
                   $= -6 - 7 = -13$

93. $-3 - \dfrac{1}{2} = \dfrac{-3}{1} - \dfrac{1}{2} = \dfrac{-3 \cdot 2}{1 \cdot 2} - \dfrac{1}{2} = \dfrac{-6}{2} - \dfrac{1}{2} = -\dfrac{7}{2}$

95. $\dfrac{4}{5} + \dfrac{1}{10} + \dfrac{3}{8} = \dfrac{4 \cdot 8}{5 \cdot 8} + \dfrac{1 \cdot 4}{10 \cdot 4} + \dfrac{3 \cdot 5}{8 \cdot 5} = \dfrac{32}{40} + \dfrac{4}{40} + \dfrac{15}{40} = \dfrac{51}{40}$

2.2 Addition Property of Equality

1. $x - 3 = 8$
   $x - 3 + 3 = 8 + 3$
   $x = 11$

3. $x + 2 = 6$
   $x + 2 + (-2) = 6 + (-2)$
   $x = 4$

5. $a + \dfrac{1}{2} = -\dfrac{1}{4}$
   $a + \dfrac{1}{2} + \left(-\dfrac{1}{2}\right) = -\dfrac{1}{4} + \left(-\dfrac{1}{2}\right)$
   $a = -\dfrac{1}{4} + \left(-\dfrac{2}{4}\right)$
   $a = -\dfrac{3}{4}$

7. $x + 2.3 = -3.5$
   $x + 2.3 + (-2.3) = -3.5 + (-2.3)$
   $x = -5.8$

9. $y + 11 = -6$
   $y + 11 + (-11) = -6 + (-11)$
   $y = -17$

11. $x - \dfrac{5}{8} = -\dfrac{3}{4}$
    $x - \dfrac{5}{8} + \dfrac{5}{8} = -\dfrac{3}{4} + \dfrac{5}{8}$
    $x = -\dfrac{6}{8} + \dfrac{5}{8}$
    $x = -\dfrac{1}{8}$

13. $m - 6 = -10$
    $m - 6 + 6 = -10 + 6$
    $m = -4$

15. $6.9 + x = 3.3$
    $-6.9 + 6.9 + x = -6.9 + 3.3$
    $x = -3.6$

17. $5 = a + 4$
    $5 + (-4) = a + 4 + (-4)$
    $a = 1$

19. $-\dfrac{5}{9} = x - \dfrac{2}{5}$
    $-\dfrac{5}{9} + \dfrac{2}{5} = x - \dfrac{2}{5} + \dfrac{2}{5}$
    $-\dfrac{25}{45} + \dfrac{18}{45} = x$
    $x = -\dfrac{7}{45}$

© 2000 Harcourt, Inc

21. $4x + 2 - 3x = 4 + 1$
$x + 2 = 5$
$x + 2 + (2) = 5 + (-2)$
$x = 3$

23. $8a - \frac{1}{2} - 7a = \frac{3}{4} + \frac{1}{8}$
$a - \frac{1}{2} = \frac{6}{8} + \frac{1}{8}$
$a - \frac{1}{2} = \frac{7}{8}$
$a - \frac{1}{2} + \frac{1}{2} = \frac{7}{8} + \frac{1}{2}$
$a = \frac{7}{8} + \frac{4}{8}$
$a = \frac{11}{8}$

25. $-3 - 4x + 5x = 18$
$-3 + x = 18$
$3 - 3 + x = 3 + 18$
$x = 21$

27. $-11x + 2 + 10x + 2x = 9$
$x + 2 = 9$
$x + 2 + (-2) = 9 + \{-2\}$
$x = 7$

29. $-2.5 + 4.8 = 8x - 1.2 - 7x$
$2.3 = x - 1.2$
$2.3 + 1.2 = x - 1.2 + 1.2$
$x = 3.5$

31. $2y - 10 + 3y - 4y = 18 - 6$
$y - 10 = 12$
$y - 10 + 10 = 12 + 10$
$y = 22$

33. $15 - 21 = 8x + 3x - 10x$
$x = -6$

35. $24 - 3 + 8a - 5a - 2a = 21$
$21 + a = 21$
$-21 + 21 + a = -21 + 21$
$a = 0$

37. $2(x + 3) - x = 4$
$2x + 6 - x = 4$
$x + 6 = 4$
$x + 6 + (-6) = 4 + (-6)$
$x = -2$

39. $-3(x - 4) + 4x = 3 - 7$
$-3x + 12 + 4x = -4$
$x + 12 = -4$
$x + 12 + (-12) = -4 + (-12)$
$x = -16$

41. $5(2a + 1) - 9a = 8 - 6$
$10a + 5 - 9a = 2$
$a + 5 = 2$
$a + 5 + (-5) = 2 + (-5)$
$a = -3$

43. $-(x + 3) + 2x - 1 = 6$
$-x - 3 + 2x - 1 = 6$
$x - 4 = 6$
$x - 4 + 4 = 6 + 4$
$x = 10$

Problem Set 2.2

45. $4y - 3(y-6) + 2 = 8$
$4y - 3y + 18 + 2 = 8$
$y + 20 = 8$
$y + 20 + (-20) = 8 + (-20)$
$y = -12$

47. $2(3x+1) - 5(x+2) = 1 - 10$
$6x + 2 - 5x - 10 = -9$
$x - 8 = -9$
$x - 8 + 8 = -9 + 8$
$x = -1$

49. $-3(2m-9) + 7(m-4) = 12 - 9$
$-6m + 27 + 7m - 28 = 3$
$m - 1 = 3$
$m - 1 + 1 = 3 + 1$
$m = 4$

51. $4x = 3x + 2$
$4x + (-3x) = 3x + (-3x) + 2$
$x = 2$

53. $8a = 7a - 5$
$8a + (-7a) = 7a + (-7a) - 5$
$a = -5$

55. $2x = 3x + 1$
$(-2x) + 2x = (-2x) + 3x + 1$
$0 = x + 1$
$0 + (-1) = x + 1 + (-1)$
$x = -1$

57. $3y + 4 = 2y + 1$
$3y + (-2y) + 4 = 2y + (-2y) + 1$
$y + 4 + (-4) = 1 + (-4)$
$y = -3$

59. $2m - 3 = m + 5$
$2m + (-m) - 3 = m + (-m) + 5$
$m - 3 = 5$
$m - 3 + 3 = 5 + 3$
$m = 8$

61. $4x - 7 = 5x + 1$
$4x + (-4x) - 7 = 5x + (-4x) + 1$
$-7 = x + 1$
$-7 + (-1) = x + 1 + (-1)$
$x = -8$

63. $5x - \dfrac{2}{3} = 4x + \dfrac{4}{3}$
$5x + (-4x) - \dfrac{2}{3} = 4x + (-4x) + \dfrac{4}{3}$
$x - \dfrac{2}{3} = \dfrac{4}{3}$
$x - \dfrac{2}{3} + \dfrac{2}{3} = \dfrac{4}{3} + \dfrac{2}{3}$
$x = \dfrac{6}{3} = 2$

© 2000 Harcourt, Inc

65.
$$8a - 7.1 = 7a + 3.9$$
$$8a + (-7a) - 7.1 = 7a + (-7a) + 3.9$$
$$a - 7.1 = 3.9$$
$$a - 7.1 + 7.1 = 3.9 + 7.1$$
$$a = 11$$

67.
$$x - 2 + 5 = 6$$
$$x + 3 = 6$$
$$x = 3$$
$$y + 6 + 1 = 6$$
$$y + 7 = 6$$
$$y = -1$$
$$4 + 2 + z = 6$$
$$6 + z = 6$$
$$z = 0$$
$$x = 3, y = -1, z = 0$$

69. $T + R + A = 100$
(a) $T = 88, A = 6$
$$88 + R + 6 = 100$$
$$R + 94 = 100$$
$$R = 6\%$$
(b) $T = 0, A = 95$
$$0 + R + 95 = 100$$
$$R = 5\%$$
(c) $T = 0, R = 98$
$$0 + 98 + A = 100$$
$$A = 2\%$$
(d) $T = 0, A = 25$
$$0 + R + 25 = 100$$
$$R = 75\%$$

71.
$$x + 55 + 55 = 180$$
$$x + 110 = 180$$
$$x = 70°$$

73.
$$x - 3 = 8$$
$$x = 11$$

75.
$$2y + 3 = y + 5$$
$$2y + (-y) + 3 = y + (-y) + 5$$
$$y + 3 = 5$$
$$y = 2$$

77. $3(6x) = (3 \cdot 6)x = 18x$

79. $\frac{1}{5}(5x) = \left(\frac{1}{5} \cdot 5\right)x = 1x = x$

81. $8\left(\frac{1}{8}y\right) = \left(8 \cdot \frac{1}{8}\right)y = 1y = y$

83. $-2\left(-\frac{1}{2}x\right) = \left[-2 \cdot \left(-\frac{1}{2}\right)\right]x = 1x = x$

Problem Set 2.3

85. $-\dfrac{4}{3}\left(-\dfrac{3}{4}a\right) = \left[-\dfrac{4}{3}\cdot\left(-\dfrac{3}{4}\right)\right]a = 1a = a$

2.3 Multiplication Property of Equality

1. $5x = 10$
$\dfrac{1}{5}(5x) = \dfrac{1}{5}(10)$
$x = 2$

3. $7a = 28$
$\dfrac{1}{7}(7a) = \dfrac{1}{7}(28)$
$a = 4$

5. $-8x = 4$
$-\dfrac{1}{8}(-8x) = -\dfrac{1}{8}(4)$
$x = -\dfrac{1}{2}$

7. $8m = -16$
$\dfrac{1}{8}(8m) = \dfrac{1}{8}(-16)$
$m = -2$

9. $-3x = -9$
$-\dfrac{1}{3}(-3x) = -\dfrac{1}{3}(-9)$
$x = 3$

11. $-7y = -28$
$-\dfrac{1}{7}(-7y) = -\dfrac{1}{7}(-28)$
$y = 4$

13. $2x = 0$
$\dfrac{1}{2}(2x) = \dfrac{1}{2}(0)$
$x = 0$

15. $-5x = 0$
$-\dfrac{1}{5}(-5x) = -\dfrac{1}{5}(0)$
$x = 0$

17. $\dfrac{x}{3} = 2$
$3\left(\dfrac{x}{3}\right) = 3(2)$
$x = 6$

19. $-\dfrac{m}{5} = 10$
$-5\left(-\dfrac{m}{5}\right) = -5(10)$
$m = -50$

21. $-\dfrac{x}{2} = -\dfrac{3}{4}$
$-2\left(-\dfrac{x}{2}\right) = -2\left(-\dfrac{3}{4}\right)$
$x = \dfrac{3}{2}$

23. $\dfrac{2}{3}a = 8$
$\dfrac{3}{2}\left(\dfrac{2}{3}a\right) = \dfrac{3}{2}(8)$
$a = 12$

© 2000 Harcourt, Inc

25. $$-\frac{3}{5}x = \frac{9}{5}$$
$$-\frac{5}{3}\left(-\frac{3}{5}x\right) = -\frac{5}{3}\left(\frac{9}{5}\right)$$
$$x = -3$$

27. $$-\frac{5}{8}y = -20$$
$$-\frac{8}{5}\left(-\frac{5}{8}y\right) = -\frac{8}{5}(-20)$$
$$y = 32$$

29. $$-4x - 2x + 3x = 24$$
$$-3x = 24$$
$$-\frac{1}{3}(-3x) = -\frac{1}{3}(24)$$
$$x = -8$$

31. $$4x + 8x - 2x = 15 - 10$$
$$10x = 5$$
$$\frac{1}{10}(10x) = \frac{1}{10}(5)$$
$$x = \frac{1}{2}$$

33. $$-3 - 5 = 3x + 5x - 10x$$
$$-\frac{1}{2}(-8) = -\frac{1}{2}(-2x)$$
$$x = 4$$

35. $$18 - 13 = \frac{1}{2}a + \frac{3}{4}a - \frac{5}{8}a$$
$$8(5) = 8\left(\frac{1}{2}a + \frac{3}{4}a - \frac{5}{8}a\right)$$
$$40 = 4a + 6a - 5a$$
$$40 = 5a$$
$$\frac{1}{5}(40) = \frac{1}{5}(5a)$$
$$a = 8$$

37. $$-x = 4$$
$$-1(-x) = -1(4)$$
$$x = -4$$

39. $$-x = -4$$
$$-1(-x) = -1(-4)$$
$$x = 4$$

41. $$15 = -a$$
$$-1(15) = -1(-a)$$
$$a = -15$$

43. $$-y = \frac{1}{2}$$
$$-1(-y) = -1\left(\frac{1}{2}\right)$$
$$y = -\frac{1}{2}$$

45. $$3x - 2 = 7$$
$$3x - 2 + 2 = 7 + 2$$
$$3x = 9$$
$$\frac{1}{3}(3x) = \frac{1}{3}(9)$$
$$x = 3$$

47. $$2a + 1 = 3$$
$$2a + 1 + (-1) = 3 + (-1)$$
$$\frac{1}{2}(2a) = \frac{1}{2}(2)$$
$$a = 1$$

© 2000 Harcourt, Inc

Problem Set 2.3

49. $\dfrac{1}{8} + \dfrac{1}{2}x = \dfrac{1}{4}$

$8\left(\dfrac{1}{8} + \dfrac{1}{2}x\right) = 8\left(\dfrac{1}{4}\right)$

$1 + 4x = 2$

$(-1) + 1 + 4x = (-1) + 2$

$4x = 1$

$\dfrac{1}{4}(4x) = \dfrac{1}{4}(1)$

$x = \dfrac{1}{4}$

51. $6x = 2x - 12$

$6x + (-2x) = 2x + (-2x) - 12$

$4x = -12$

$\dfrac{1}{4}(4x) = \dfrac{1}{4}(-12)$

$x = -3$

53. $2y = -4y + 18$

$2y + 4y = -4y + 4y + 18$

$6y = 18$

$\dfrac{1}{6}(6y) = \dfrac{1}{6}(18)$

$y = 3$

55. $-7x = -3x - 8$

$-7x + 3x = -3x + 3x - 8$

$-4x = -8$

$-\dfrac{1}{4}(-4x) = -\dfrac{1}{4}(-8)$

$x = 2$

57. $8x + 4 = 2x - 5$

$8x + (-2x) + 4 = 2x + (-2x) - 5$

$6x + 4 = -5$

$6x + 4 + (-4) = -5 + (-4)$

$6x = -9$

$\dfrac{1}{6}(6x) = \dfrac{1}{6}(-9)$

$x = -\dfrac{3}{2}$

59. $x + \dfrac{1}{2} = \dfrac{1}{4}x - \dfrac{5}{8}$

$8\left(x + \dfrac{1}{2}\right) = 8\left(\dfrac{1}{4}x - \dfrac{5}{8}\right)$

$8x + 4 = 2x - 5$

$8x + (-2x) + 4 = 2x + (-2x) - 5$

$6x + 4 = -5$

$6x + 4 + (-4) = -5 + (-4)$

$6x = -9$

$\dfrac{1}{6}(6x) = \dfrac{1}{6}(-9)$

$x = -\dfrac{3}{2}$

61. $6m - 3 = m + 2$

$6m + (-m) - 3 = m + (-m) + 2$

$5m - 3 = 2$

$5m - 3 + 3 = 2 + 3$

$5m = 5$

$\dfrac{1}{5}(5m) = \dfrac{1}{5}(5)$

$m = 1$

© 2000 Harcourt, Inc

63.
$$\frac{1}{2}m - \frac{1}{4} = \frac{1}{12}m + \frac{1}{6}$$
$$12\left(\frac{1}{2}m - \frac{1}{4}\right) = 12\left(\frac{1}{12}m + \frac{1}{6}\right)$$
$$6m - 3 = m + 2$$
$$6m + (-m) - 3 = m + (-m) + 2$$
$$5m - 3 = 2$$
$$5m - 3 + 3 = 2 + 3$$
$$5m = 5$$
$$\frac{1}{5}(5m) = \frac{1}{5}(5)$$
$$m = 1$$

65.
$$9y + 2 = 6y - 4$$
$$9y + (-6y) + 2 = 6y + (-6y) - 4$$
$$3y + 2 = -4$$
$$3y + 2 + (-2) = -4 + (-2)$$
$$3y = -6$$
$$\frac{1}{3}(3y) = \frac{1}{3}(-6)$$
$$y = -2$$

67.
$$\frac{3}{2}y + \frac{1}{3} = y - \frac{2}{3}$$
$$6\left(\frac{3}{2}y + \frac{1}{3}\right) = 6\left(y - \frac{2}{3}\right)$$
$$9y + 2 = 6y - 4$$
$$9y + (-6y) + 2 = 6y + (-6y) - 4$$
$$3y + 2 = -4$$
$$3y + 2 + (-2) = -4 + (-2)$$
$$3y = -6$$
$$\frac{1}{3}(3y) = \frac{1}{3}(-6)$$
$$y = -2$$

69.
$$7.5x = 1500$$
$$\frac{1}{7.5}(7.5x) = \frac{1}{7.5}(1500)$$
$$x = 200$$

71.
$$1 + 3(2) + 3x = 13$$
$$1 + 6 + 3x = 13$$
$$7 + 3x = 13$$
$$7 + (-7) + 3x = 13 + (-7)$$
$$3x = 6$$
$$\frac{1}{3}(3x) = \frac{1}{3}(6)$$
$$x = 2$$

73. (a) $AB = 12$, $AY = 15$, and $AC = 20$
$$\frac{AX}{12} = \frac{15}{20}$$
$$12\left(\frac{AX}{12}\right) = 12\left(\frac{15}{20}\right)$$
$$AX = 9$$
(b) $XY = 8$, $BC = 10$, and $AX = 12$
$$\frac{12}{AB} = \frac{8}{10}$$
$$AB\left(\frac{12}{AB}\right) = AB\left(\frac{8}{10}\right)$$
$$12 = \frac{8}{10}AB$$
$$\frac{10}{8}(12) = \frac{10}{8}\left(\frac{8}{10}AB\right)$$
$$15 = AB$$

Problem Set 2.3

73.

Continued

(c) $YC = 6$, $XB = 4$, and $AX = 8$

$$\frac{8}{8+4} = \frac{AY}{AY+6}$$

$$12(AY+6)\left(\frac{8}{12}\right) = 12(AY+6)\left(\frac{AY}{AY+6}\right)$$

$$8(AY+6) = 12AY$$

$$8AY + 48 = 12AY$$

$$8AY + (-8AY) + 48 = 12AY + (-8AY)$$

$$48 = 4AY$$

$$\frac{1}{4}(48) = \frac{1}{4}(4AY)$$

$$12 = AY$$

$$AC = AY + YC = 12 + 6 = 18$$

75.

$$3x + 2 = 19$$

$$3x + 2 + (-2) = 19 + (-2)$$

$$3x = 17$$

$$\frac{1}{3}(3x) = \frac{1}{3}(17)$$

$$x = \frac{17}{3}$$

77.

$$2(x+10) = 40$$

$$\frac{1}{2}(2)(x+10) = \frac{1}{2}(40)$$

$$x + 10 = 20$$

$$x + 10 + (-10) = 20 + (-10)$$

$$x = 10$$

79. Using the distributive property and combining like terms: $5(2x-8) - 3 = 10x - 40 - 3 = 10x - 43$

81. Using the distributive property and combining like terms:

$-2(3x+5) + 3(x-1) = -6x - 10 + 3x - 3 = -6x + 3x - 10 - 3 = -3x - 13$

83. Using the distributive property and combining like terms: $7 - 3(2y+1) = 7 - 6y - 3 = -6y + 7 - 3 = -6y + 4$

85. Using the distributive property and combining like terms: $4x - (9x-3) + 4 = 4x - 9x + 3 + 4 = -5x + 7$

© 2000 Harcourt, Inc

2.4 Solving Linear Equations

1. $2(x+3) = 12$
$2x+6 = 12$
$2x+6+(-6) = 12+(-6)$
$2x = 6$
$\frac{1}{2}(2x) = \frac{1}{2}(6)$
$x = 3$

3. $6(x-1) = -18$
$6x-6 = -18$
$6x-6+6 = -18+6$
$6x = -12$
$\frac{1}{6}(6x) = \frac{1}{6}(-12)$
$x = -2$

5. $2(4a+1) = -6$
$8a+2 = -6$
$8a+2+(-2) = -6+(-2)$
$8a = -8$
$\frac{1}{8}(8a) = \frac{1}{8}(-8)$
$a = -1$

7. $14 = 2(5x-3)$
$14 = 10x-6$
$14+6 = 10x-6+6$
$20 = 10x$
$\frac{1}{10}(20) = \frac{1}{10}(10x)$
$x = 2$

9. $-2(3y+5) = 14$
$-6y-10 = 14$
$-6y-10+10 = 14+10$
$-6y = 24$
$-\frac{1}{6}(-6y) = -\frac{1}{6}(24)$
$y = -4$

11. $-5(2a+4) = 0$
$-10a-20 = 0$
$-10a-20+20 = 0+20$
$-10a = 20$
$-\frac{1}{10}(-10a) = -\frac{1}{10}(20)$
$a = -2$

13. $1 = \frac{1}{2}(4x+2)$
$1 = 2x+1$
$1+(-1) = 2x+1+(-1)$
$0 = 2x$
$\frac{1}{2}(0) = \frac{1}{2}(2x)$
$x = 0$

15. $3(t-4)+5 = -4$
$3t-12+5 = -4$
$3t-7 = -4$
$3t-7+7 = -4+7$
$3t = 3$
$\frac{1}{3}(3t) = \frac{1}{3}(3)$
$t = 1$

Problem Set 2.4

17. $4(2y+1)-7=1$
$8y+4-7=1$
$8y-3=1$
$8y-3+3=1+3$
$8y=4$
$\frac{1}{8}(8y)=\frac{1}{8}(4)$
$y=\frac{1}{2}$

19. $\frac{1}{2}(x-3)=\frac{1}{4}(x+1)$
$\frac{1}{2}x-\frac{3}{2}=\frac{1}{4}x+\frac{1}{4}$
$4\left(\frac{1}{2}x-\frac{3}{2}\right)=4\left(\frac{1}{4}x+\frac{1}{4}\right)$
$2x-6=x+1$
$2x+(-x)-6=x+(-x)+1$
$x-6=1$
$x-6+6=1+6$
$x=7$

21. $-0.7(2x-7)=0.3(11-4x)$
$-1.4x+4.9=3.3-1.2x$
$-1.4x+1.2x+4.9=3.3-1.2x+1.2x$
$-0.2x+4.9=3.3$
$-0.2x+4.9+(-4.9)=3.3+(-4.9)$
$-0.2x=-1.6$
$\frac{-0.2x}{-0.2}=\frac{-1.6}{-0.2}$
$x=8$

23. $-2(3y+1)=3(1-6y)-9$
$-6y-2=3-18y-9$
$-6y-2=-18y-6$
$-6y+18y-2=-18y+18y-6$
$12y-2=-6$
$12y-2+2=-6+2$
$12y=-4$
$\frac{1}{12}(12y)=\frac{1}{12}(-4)$
$y=-\frac{1}{3}$

25. $\frac{3}{4}(8x-4)+3=\frac{2}{5}(5x+10)-1$
$6x-3+3=2x+4-1$
$6x=2x+3$
$6x+(-2x)=2x+(-2x)+3$
$4x=3$
$\frac{1}{4}(4x)=\frac{1}{4}(3)$
$x=\frac{3}{4}$

27. $0.06x+0.08(100-x)=6.5$
$0.06x+8-0.08x=6.5$
$-0.02x+8=6.5$
$-0.02x+8+(-8)=6.5+(-8)$
$-0.02x=-1.5$
$\frac{-0.02x}{-0.02}=\frac{-1.5}{-0.02}$
$x=75$

© 2000 Harcourt, Inc

29.
$$6-5(2a-3)=1$$
$$6-10a+15=1$$
$$-10a+21=1$$
$$-10a+21+(-21)=1+(-21)$$
$$-10a=-20$$
$$-\frac{1}{10}(-10a)=-\frac{1}{10}(-20)$$
$$a=2$$

31.
$$0.2x-0.5=0.5-0.2(2x-13)$$
$$0.2x-0.5=0.5-0.4x+2.6$$
$$0.2x-0.5=-0.4x+3.1$$
$$0.2x+0.4x-0.5=-0.4x+0.4x+3.1$$
$$0.6x-0.5=3.1$$
$$0.6x-0.5+0.5=3.1+0.5$$
$$0.6x=3.6$$
$$\frac{0.6x}{0.6}=\frac{3.6}{0.6}$$
$$x=6$$

33.
$$2(t-3)+3(t-2)=28$$
$$2t-6+3t-6=28$$
$$5t-12=28$$
$$5t-12+12=28+12$$
$$5t=40$$
$$\frac{1}{5}(5t)=\frac{1}{5}(40)$$
$$t=8$$

35.
$$5(x-2)-(3x+4)=3(6x-8)+10$$
$$5x-10-3x-4=18x-24+10$$
$$2x-14=18x-14$$
$$2x+(-18x)-14=18x+(-18x)-14$$
$$-16x-14=-14$$
$$-16x-14+14=-14+14$$
$$-16x=0$$
$$-\frac{1}{16}(-16x)=-\frac{1}{16}(0)$$
$$x=0$$

37.
$$2(5x-3)-(2x-4)=5-(6x+1)$$
$$10x-6-2x+4=5-6x-1$$
$$8x-2=-6x+4$$
$$8x+6x-2=-6x+6x+4$$
$$14x-2+2=4+2$$
$$14x=6$$
$$\frac{1}{14}(14x)=\frac{1}{14}(6)$$
$$x=\frac{3}{7}$$

39.
$$-(3x+1)-(4x-7)=4-(3x+2)$$
$$-3x-1-4x+7=4-3x-2$$
$$-7x+6=-3x+2$$
$$-7x+3x+6=-3x+3x+2$$
$$-4x+6=2$$
$$-4x+6+(-6)=2+(-6)$$
$$-4x=-4$$
$$-\frac{1}{4}(-4x)=-\frac{1}{4}(-4)$$
$$x=1$$

41. $\frac{1}{2}(3) = \frac{1}{2} \cdot \frac{3}{1} = \frac{3}{2}$

43. $\frac{2}{3}(6) = \frac{2}{3} \cdot \frac{6}{1} = \frac{12}{3} = 4$

45. $\frac{5}{9} \cdot \frac{9}{5} = \frac{45}{45} = 1$

47. $2(3x-5) = 2 \cdot 3x - 2 \cdot 5 = 6x - 10$

49. $\frac{1}{2}(3x+6) = \frac{1}{2} \cdot (3x) + \frac{1}{2} \cdot 6 = \frac{3}{2}x + 3$

51. $\frac{1}{3}(-3x+6) = \frac{1}{3} \cdot (-3x) + \frac{1}{3} \cdot 6 = -x + 2$

Problem Set 2.5

2.5 Formulas

1. Substituting $P = 300$ and $w = 50$:
$$P = 2l + 2w$$
$$300 = 2l + 2(50)$$
$$300 = 2l + 100$$
$$200 = 2l$$
$$l = 100$$

3. Substituting $x = 3$:
$$2(3) + 3y = 6$$
$$6 + 3y = 6$$
$$6 + (-6) + 3y = 6 + (-6)$$
$$3y = 0$$
$$y = 0$$

5. Substituting $x = 0$:
$$2(0) + 3y = 6$$
$$0 + 3y = 6$$
$$3y = 6$$
$$y = 2$$

7. Substituting $y = 2$:
$$2x - 5(2) = 20$$
$$2x - 10 = 20$$
$$2x - 10 + 10 = 20 + 10$$
$$2x = 30$$
$$x = 15$$

9. Substituting $y = 0$:
$$2x - 5(0) = 20$$
$$2x - 0 = 20$$
$$2x = 20$$
$$x = 10$$

11. Substituting $y = 7$:
$$7 = 2x - 1$$
$$7 + 1 = 2x - 1 + 1$$
$$2x = 8$$
$$x = 4$$

13. Substituting $y = 3$:
$$3 = 2x - 1$$
$$3 + 1 = 2x - 1 + 1$$
$$2x = 4$$
$$x = 2$$

15. Solving for l:
$$lw = A$$
$$\frac{lw}{w} = \frac{A}{w}$$
$$l = \frac{A}{w}$$

17. Solving for r:
$$rt = d$$
$$\frac{rt}{t} = \frac{d}{t}$$
$$r = \frac{d}{t}$$

19. Solving for h:
$$lwh = V$$
$$\frac{lwh}{lw} = \frac{V}{lw}$$
$$h = \frac{V}{lw}$$

© 2000 Harcourt, Inc

21. Solving for P:
$$PV = nRT$$
$$\frac{PV}{V} = \frac{nRT}{V}$$
$$P = \frac{nRT}{V}$$

23. Solving for a:
$$a+b+c = P$$
$$a+b+c-b-c = P-b-c$$
$$a = P-b-c$$

25. Solving for x:
$$x-3y = -1$$
$$x-3y+3y = -1+3y$$
$$x = 3y-1$$

27. Solving for y:
$$-3x+y = 6$$
$$-3x+3x+y = 6+3x$$
$$y = 3x+6$$

29. Solving for y:
$$2x+3y = 6$$
$$-2x+2x+3y = -2x+6$$
$$3y = -2x+6$$
$$\frac{1}{3}(3y) = \frac{1}{3}(-2x+6)$$
$$y = -\frac{2}{3}x+2$$

31. Solving for y:
$$6x+3y = 12$$
$$-6x+6x+3y = -6x+12$$
$$3y = -6x+12$$
$$\frac{1}{3}(3y) = \frac{1}{3}(-6x+12)$$
$$y = -2x+4$$

33. Solving for y:
$$5x-2y = 3$$
$$-5x+5x-2y = -5x+3$$
$$-2y = -5x+3$$
$$-\frac{1}{2}(-2y) = -\frac{1}{2}(-5x+3)$$
$$y = \frac{5}{2}x - \frac{3}{2}$$

35. Solving for w:
$$2l+2w = P$$
$$2l-2l+2w = P-2l$$
$$2w = P-2l$$
$$\frac{2w}{2} = \frac{P-2l}{2}$$
$$w = \frac{P-2l}{2}$$

37. Solving for v:
$$vt+16t^2 = h$$
$$vt+16t^2-16t^2 = h-16t^2$$
$$vt = h-16t^2$$
$$\frac{vt}{t} = \frac{h-16t^2}{t}$$
$$v = \frac{h-16t^2}{t}$$

39. Solving for h:
$$\pi r^2 + 2\pi rh = A$$
$$\pi r^2 - \pi r^2 + 2\pi rh = A - \pi r^2$$
$$2\pi rh = A - \pi r^2$$
$$\frac{2\pi rh}{2\pi r} = \frac{A-\pi r^2}{2\pi r}$$
$$h = \frac{A-\pi r^2}{2\pi r}$$

Problem Set 2.5

41. $\dfrac{x}{2} + \dfrac{y}{3} = 1$

$-\dfrac{x}{2} + \dfrac{x}{2} + \dfrac{y}{3} = -\dfrac{x}{2} + 1$

$\dfrac{y}{3} = -\dfrac{x}{2} + 1$

$3\left(\dfrac{y}{3}\right) = 3\left(-\dfrac{x}{2} + 1\right)$

$y = -\dfrac{3}{2}x + 3$

43. $\dfrac{x}{7} - \dfrac{y}{3} = 1$

$-\dfrac{x}{7} + \dfrac{x}{7} - \dfrac{y}{3} = -\dfrac{x}{7} + 1$

$-\dfrac{y}{3} = -\dfrac{x}{7} + 1$

$-3\left(-\dfrac{y}{3}\right) = -3\left(-\dfrac{x}{7} + 1\right)$

$y = \dfrac{3}{7}x - 3$

45. $-\dfrac{1}{4}x + \dfrac{1}{8}y = 1$

$-\dfrac{1}{4}x + \dfrac{1}{4}x + \dfrac{1}{8}y = 1 + \dfrac{1}{4}x$

$\dfrac{1}{8}y = \dfrac{1}{4}x + 1$

$8\left(\dfrac{1}{8}y\right) = 8\left(\dfrac{1}{4}x + 1\right)$

$y = 2x + 8$

47. The complement of $30°$ is $90° - 30° = 60°$, and the supplement is $180° - 30° = 150°$.

49. The complement of $45°$ is $90° - 45° = 45°$, and the supplement is $180° - 45° = 135°$.

51. $x = 0.25 \cdot 40$
$x = 10$
The number 10 is 25% of 40.

53. $x = 0.12 \cdot 2000$
$x = 240$
The number 240 is 12% of 2000.

55. $x \cdot 28 = 7$
$8x = 7$
$\dfrac{1}{28}(28x) = \dfrac{1}{28}(7)$
$x = 0.25 = 25\%$
The number 7 is 25% of 28.

57. $x \cdot 40 = 14$
$40x = 14$
$\dfrac{1}{40}(40x) = \dfrac{1}{40}(14)$
$x = 0.35 = 35\%$
The number 14 is 35% of 40.

59. $0.50 \cdot x = 32$
$\dfrac{0.50x}{0.50} = \dfrac{32}{0.50}$
$x = 64$
The number 32 is 50% of 64.

61. $0.12 \cdot x = 240$
$\dfrac{0.12x}{0.12} = \dfrac{240}{0.12}$
$x = 2000$
The number 240 is 12% of 2000.

© 2000 Harcourt, Inc

63. (a) 62.8% of T is $68,840,000$
$$0.628T = 68,840,000$$
(b) $$\frac{0.628T}{0.628} = \frac{68,840,000}{0.628}$$
$$T = 109,617,834 \text{ people}$$

65. (a) 12 is what percent of 20?
$$12 = x \cdot 20$$
$$x = \frac{12}{20} = 0.6 = 60\% \text{ silver}$$
(b) $100\% - 60\% = 40\%$ copper

67. Substituting $F = 212$:
$$C = \frac{5}{9}(212 - 32)$$
$$= \frac{5}{9}(180)$$
$$= 100°C$$
Yes

69. Substituting $F = 68$:
$$C = \frac{5}{9}(68 - 32)$$
$$= \frac{5}{9}(36)$$
$$= 20°$$
Yes

71. $$\frac{9}{5}C + 32 = F$$
$$\frac{9}{5}C + 32 - 32 = F - 32$$
$$\frac{9}{5}C = F - 32$$
$$\frac{5}{9}\left(\frac{9}{5}C\right) = \frac{5}{9}(F - 32)$$
$$C = \frac{5}{9}(F - 32)$$

73. Find what percent of 150 is 90:
$$x \cdot 150 = 90$$
$$\frac{1}{150}(150x) = \frac{1}{150}(90)$$
$$x = 0.60 = 60\%$$
60% of the calories in one serving of vanilla ice cream are fat calories.

75. Find what percent of 98 is 26:
$$x \cdot 98 = 26$$
$$\frac{1}{98}(98x) = \frac{1}{98}(26)$$
$$x \approx 0.265 = 26.5\%$$
26.5% of one serving of frozen yogurt are carbohydrates.

77. $T = -0.0035A + 56$
See the table in the back of the textbook.

Problem Set 2.6

79. (a)   If $C = 44$ and $\pi = \dfrac{22}{7}$

$$2 \cdot \dfrac{22}{7} r = 44$$
$$\dfrac{44}{7} r = 44$$
$$\dfrac{7}{44}\left(\dfrac{44}{7} r\right) = \dfrac{7}{44}(44)$$
$$r = 7 \text{ meters}$$

(b)   If $C = 9.42$ and $\pi = 3.14$

$$2 \cdot (3.14) r = 9.42$$
$$6.28 r = 9.42$$
$$\dfrac{6.28 r}{6.28} = \dfrac{9.42}{6.28}$$
$$r = 1.5 \text{ inches}$$

81. The sum of 4 and 1 is 5.

83. The difference of 6 and 2 is 4.

85. An equivalent expression is: $2(6+3)$

87. An equivalent expression is: $2(5) + 3$

2.6 Applications

1. Let $x =$ the number.
$$x + 5 = 13$$
$$x = 8$$
The number is 8.

3. Let $x =$ the number.
$$2x + 4 = 14$$
$$2x = 10$$
$$x = 5$$
The number is 5.

5. Let $x =$ the number.
$$5(x+7) = 30$$
$$5x + 35 = 30$$
$$5x = -5$$
$$x = -1$$
The number is -1.

7. Let x and $x + 2 =$ the two numbers.
$$x + x + 2 = 8$$
$$2x + 2 = 8$$
$$2x = 6$$
$$x = 3$$
$$x + 2 = 5$$
The two numbers are 3 and 5.

9. Let x and $3x-4 =$ the two numbers.

$$(x+3x-4)+5 = 25$$
$$4x+1 = 25$$
$$4x = 24$$
$$x = 6$$
$$3x-4 = 3(6)-4 = 14$$

The two numbers are 6 and 14.

11.

	Five Years Ago	Now
Fred	$x+4-5 = x-1$	$x+4$
Barney	$x-5$	x

$$x-1+x-5 = 48$$
$$2x-6 = 48$$
$$2x = 54$$
$$x = 27$$
$$x+4 = 31$$

Barney is 27 and Fred is 31.

13.

	Now	Three years from now
Jack	$2x$	$2x+3$
Lacy	x	$x+3$

$$2x+3+x+3 = 54$$
$$3x+6 = 54$$
$$3x = 48$$
$$x = 16$$
$$2x = 32$$

Lacy is 16 and Jack is 32.

15.

	Now	Two years from now
Pat	$x+20$	$x+20+2 = x+22$
Patrick	x	$x+2$

$$x+22 = 2(x+2)$$
$$x+22 = 2x+4$$
$$2 = x+4$$
$$x = 18$$
$$20 = 38$$

Patrick is 18 and Pat is 38.

17. Let $w =$ the width and $w+5 =$ the length

$$2w+2(w+5) = 34$$
$$2w+2w+10 = 34$$
$$4w+10 = 34$$
$$4x = 24$$
$$w = 6$$
$$w+5 = 11$$

The length is 11 inches and the width is 6 inches.

19. Let s represent the side of the square.

$$4s = 48$$
$$s = 12$$

The length of one side is 12 meters.

21. Let $w =$ the width and $2w-3 =$ the length.

$$2w+2(2w-3) = 54$$
$$2w+4w-6 = 54$$
$$6w-6 = 54$$
$$6w = 60$$
$$w = 10$$
$$2w-3 = 2(10)-3 = 17$$

The length is 17 inches and the width is 10 inches.

23.

	Nickels	Dimes
Number	x	$x+9$
Value (cents)	$5(x)$	$10(x+9)$

$$5(x)+10(x+9) = 210$$
$$x+10x+90 = 210$$
$$15x+90 = 210$$
$$15x = 120$$
$$x = 8$$
$$x+9 = 17$$

Sue has 8 nickels and 17 dimes

Problem Set 2.6

25.

	Dimes	Quarters
Number	x	$2x$
Value (cents)	$10(x)$	$25(2x)$

$$10(x) + 25(2x) = 900$$
$$10x + 50x = 900$$
$$60x = 900$$
$$x = 15$$
$$2x = 30$$

You have 15 dimes and 30 quarters.

27.

	Nickels	Dimes	Quarters
Number	x	$x+3$	$x+5$
Value (cents)	$5(x)$	$10(x+3)$	$25(x+5)$

$$5(x) + 10(x+3) + 25(x+5) = 435$$
$$5x + 10x + 30 + 25x + 125 = 435$$
$$40x + 155 = 435$$
$$40x = 280$$
$$x = 7$$
$$x + 3 = 10$$
$$x + 5 = 12$$

Katie has 7 nickels, 10 dimes, and 12 quarters.

29. 4 is less than 10

31. 9 is greater than or equal to -5

33. $12 < 20$

35. $-8 < -6$

37. $|8 - 3| - |5 - 2| = |5| - |3| = 5 - 3 = 2$

39. $15 - |9 - 3(7 - 5)| = 15 - |9 - 3(2)| = 15 - |9 - 6| = 15 - |3| = 15 - 3 = 12$

2.7 More Applications

1.

	Dollars Invested at 8%	Dollars Invested at 9%
Number of	x	$x + 2000$
Interest on	$0.08(x)$	$0.09(x+2000)$

$$0.08(x) + 0.09(x+2000) = 860$$
$$0.08x + 0.09x + 180 = 860$$
$$0.17x + 180 = 860$$
$$0.17x = 680$$
$$x = 4000$$
$$x + 2000 = 6000$$

You have $4,000 invested at 8% and $6,000 invested at 9%.

3.

	Dollars Invested at 10%	Dollars Invested at 12%
Number of	x	$x + 500$
Interest on	$0.10(x)$	$0.12(x+500)$

$$0.10(x) + 0.12(x+500) = 214$$
$$0.10x + 0.12x + 60 = 214$$
$$0.22x + 60 = 214$$
$$0.22x = 154$$
$$x = 700$$
$$x + 500 = 1200$$

Tyler has $700 invested at 10% and $1,200 invested at 12%.

5.

	Dollars Invested at 8%	Dollars Invested at 9%	Dollars Invested at 10%
Number of	x	$2x$	$3x$
Interest on	$0.08(x)$	$0.09(2x)$	$0.10(3x)$

$$0.08(x) + 0.09(2x) + 0.10(3x) = 280$$
$$0.08x + 0.18x + 0.30x = 280$$
$$0.56x = 280$$
$$x = 500$$
$$2x = 1000$$
$$3x = 1500$$

She has invested $500 at 8%, $1,000 at 9% and $1,500 at 10%.

© 2000 Harcourt, Inc

Problem Set 2.7

7. Let x represent the measure of the two equal angles, so $x + x = 2x$ represents the measure of the third angle. Since the sum of the three angles is $180°$, the equation is:

$$x + x + 2x = 180°$$
$$4x = 180°$$
$$x = 45°$$
$$x = 90°$$

The measures of the three angles are $45°$, $45°$, and $90°$.

9. Let $x =$ the largest angle. Then $\frac{1}{5}x =$ the smallest angle, and $2\left(\frac{1}{5}x\right) = \frac{2}{5}x =$ the other angle.

$$x + \frac{1}{5}x + \frac{2}{5}x = 180°$$
$$\frac{5}{5}x + \frac{1}{5}x + \frac{2}{5}x = 180°$$
$$\frac{8}{5}x = 180$$
$$x = 112.5°$$
$$\frac{1}{5}x = 22.5°$$
$$\frac{2}{5}x = 45°$$

The three angles are $22.5°, 45°, 112.5°$.

11. Let $x =$ the other acute angle, and $90°$ is the measure of the right angle. The sum of the three angles is $180°$.

$$x + 37° + 90° = 180°$$
$$x + 127° = 180°$$
$$x = 53°$$

The other two angles are $53°$ and $90°$.

13. Let $x =$ the total minutes for the call. Then $0.41 is charged for the first minute, and $0.32 is charged for the additional $x - 1$ minutes.

$$0.41(1) + 0.32(x - 1) = 5.21$$
$$0.41 + 0.32x - 0.32 = 5.21$$
$$0.32x + 0.09 = 5.21$$
$$0.32x = 5.12$$
$$x = 16$$

The call was 16 minutes long.

© 2000 Harcourt, Inc

15. Let $x =$ the hours JoAnn worked that week. Then $12/hour is paid for the first 35 hours and $18/hour is paid for the additional $x - 35$ hours.

$$12(35) + 18(x - 35) = 492$$
$$420 + 18x - 630 = 492$$
$$18x - 210 = 492$$
$$18x = 702$$
$$x = 39$$

JoAnn worked 39 hours that week.

17. Let $x =$ the number of children's tickets Stacey sold, so $2x$ represents the number of adult tickets sold.

$$6.00(2x) + 4.50(x) = 115.50$$
$$12x + 4.5x = 115.5$$
$$16.5x = 115.5$$
$$x = 7$$
$$2x = 14$$

Stacey sold 7 children's tickets and 14 adult tickets.

19. For Jeff, the total time traveled is $\dfrac{425 \text{ miles}}{55 \text{ miles/hour}} \approx 7.72$ hours ≈ 463 minutes. Since he left at 11:00 AM, he will arrive at 6:43 PM. For Carla, the total time traveled is $\dfrac{425 \text{ miles}}{65 \text{ miles/hour}} \approx 6.54$ hours ≈ 392 minutes. Since she left at 1:00 PM, she will arrive at 7:32 PM. Thus Jeff will arrive in Lake Tahoe first.

21. Since $\dfrac{1}{5}$ mile = 0.2 mile, the taxi charge is $1.25 for the first $\dfrac{1}{5}$ mile and $0.25 per fifth mile for the remaining 7.3 miles. Since 7.3 miles $= \dfrac{7.3}{0.2} = 36.5$ fifths, the total charge is: $1.25 + $0.25(36.5) \approx 10.38.

23. The first $\dfrac{1}{5}$ mile is $1.25, and the remaining $12.4 - 0.2 = 12.2$ miles will be charged at $0.25 per fifth mile. Since 12.2 miles $= \dfrac{12.2}{0.2} = 61$ fifths, the total charge is: $1.25 + $0.25(61) = 16.50.

Yes, the meter is working correctly.

Problem Set 2.7

25. If all 36 people are Elk's Lodge members (which would be the least amount), the cost of the lessons would be $3(36) = \$108$. Since half of the money is paid to Ike and Nancy, the least amount they could make is $\frac{1}{2}(\$108) = \$54$.

27. Yes. The total receipts were $160, which is possible if there were 10 Elk's members and 26 nonmembers. Computing the total receipts: $10(\$3) + 26(\$5) = \$30 + \$130 = \$160$.

29. Let $x =$ salary in 1998
$0.07x =$ salary increase
$x + 0.07x = 85,200$
$1.07x = 85,200$
$x = 79,626$
The salary in 1998 was $79,626

31. The pattern is to add -4, so the next number is:
$-4 + (-4) = -8$

33. The pattern is to multiply by $-\frac{1}{2}$, so the next number is:
$$-\frac{3}{2}\left(-\frac{1}{2}\right) = \frac{3}{4}$$

35. Each number is the square of a number, with alternating signs. For example, $1^2 = 1$, $-2^2 = -4$, $3^2 = 9$, and $-4^2 = -16$. Based on this pattern, the next number is: $5^2 = 25$.

37. The pattern is to add $-\frac{1}{2}$, so the next number is: $\frac{1}{2} + \left(-\frac{1}{2}\right) = 0$

2.8 Linear Inequalities

1. $x - 5 < 7$
   $x - 5 + 5 < 7 + 5$
   $x < 12$
   See the graph in the back of the textbook.

3. $a - 4 \leq 8$
   $a - 4 + 4 \leq 8 + 4$
   $a \leq 12$
   See the graph in the back of the textbook.

5. $x - 4.3 > 8.7$
   $x - 4.3 + 4.3 > 8.7 + 4.3$
   $x > 13$
   See the graph in the back of the textbook.

7. $y + 6 \geq 10$
   $y + 6 + (-6) \geq 10 + (-6)$
   $y \geq 4$
   See the graph in the back of the textbook.

9. $2 < x - 7$
   $2 + 7 < x - 7 + 7$
   $9 < x$
   $x > 9$
   See the graph in the back of the textbook.

11. $3x < 6$
    $\frac{1}{3}(3x) < \frac{1}{3}(6)$
    $x < 2$
    See the graph in the back of the textbook.

13. $5a \leq 25$
    $\frac{1}{5}(5a) \leq \frac{1}{5}(25)$
    $a \leq 5$
    See the graph in the back of the textbook.

15. $\frac{x}{3} > 5$
    $3\left(\frac{x}{3}\right) > 3(5)$
    $x > 15$
    See the graph in the back of the textbook.

17. $-2x > 6$
    $-\frac{1}{2}(-2x) < -\frac{1}{2}(6)$
    $x < -3$
    See the graph in the back of the textbook.

19. $-3x \geq -18$
    $-\frac{1}{3}(-3x) \leq -\frac{1}{3}(-18)$
    $x \leq 6$
    See the graph in the back of the textbook.

21. $-\frac{x}{5} \leq 10$
    $-5\left(-\frac{x}{5}\right) \geq -5(10)$
    $x \geq -50$
    See the graph in the back of the textbook.

23. $-\frac{2}{3}y > 4$
    $-\frac{3}{2}\left(-\frac{2}{3}y\right) < -\frac{3}{2}(4)$
    $y < -6$
    See the graph in the back of the textbook.

© 2000 Harcourt, Inc

Problem Set 2.8

25.
$$2x - 3 < 9$$
$$2x - 3 + 3 < 9 + 3$$
$$2x < 12$$
$$\frac{1}{2}(2x) < \frac{1}{2}(12)$$
$$x < 6$$

See the graph in the back of the textbook.

27.
$$-\frac{1}{5}y - \frac{1}{3} \leq \frac{2}{3}$$
$$-\frac{1}{5}y - \frac{1}{3} + \frac{1}{3} \leq \frac{2}{3} + \frac{1}{3}$$
$$-\frac{1}{5}y \leq 1$$
$$-5\left(-\frac{1}{5}y\right) \geq -5(1)$$
$$y \geq -5$$

See the graph in the back of the textbook.

29.
$$-4x + 1 > -11$$
$$-4x + 1 + (-1) > -11 + (-1)$$
$$-4x > -12$$
$$-\frac{1}{4}(-4x) < -\frac{1}{4}(-12)$$
$$x < 3$$

See the graph in the back of the textbook.

31.
$$\frac{2}{3}x - 5 \leq 7$$
$$\frac{2}{3}x - 5 + 5 \leq 7 + 5$$
$$\frac{2}{3}x \leq 12$$
$$\frac{3}{2}\left(\frac{2}{3}x\right) \leq \frac{3}{2}(12)$$
$$x \leq 18$$

See the graph in the back of the textbook.

33.
$$-\frac{2}{5}a - 3 > 5$$
$$-\frac{2}{5}a - 3 + 3 > 5 + 3$$
$$-\frac{2}{5}a > 8$$
$$-\frac{5}{2}\left(-\frac{2}{5}a\right) < -\frac{5}{2}(8)$$
$$a < -20$$

See the graph in the back of the textbook.

35.
$$5 - \frac{3}{5}y > -10$$
$$-5 + 5 - \frac{3}{5}y > -5 + (-10)$$
$$-\frac{3}{5}y > -15$$
$$-\frac{5}{3}\left(-\frac{3}{5}y\right) < -\frac{5}{3}(-15)$$
$$y < 25$$

See the graph in the back of the textbook.

37.
$$0.3(a + 1) \leq 1.2$$
$$0.3a + 0.3 \leq 1.2$$
$$0.3a + 0.3 + (-0.3) \leq 1.2 + (-0.3)$$
$$0.3a \leq 0.9$$
$$\frac{0.3a}{0.3} \leq \frac{0.9}{0.3}$$
$$a \leq 3$$

See the graph in the back of the textbook.

39.
$$2(5 - 2x) \leq -20$$
$$10 - 4x \leq -20$$
$$-10 + 10 - 4x \leq -10 + (-20)$$
$$-4x \leq -30$$
$$-\frac{1}{4}(-4x) \geq -\frac{1}{4}(-30)$$
$$x \geq \frac{15}{2}$$

See the graph in the back of the textbook.

© 2000 Harcourt, Inc

41.
$$3x - 5 > 8x$$
$$-3x + 3x - 5 > -3x + 8x$$
$$-5 > 5x$$
$$\frac{1}{5}(-5) > \frac{1}{5}(5x)$$
$$-1 > x$$
$$x < -1$$

See the graph in the back of the textbook.

43.
$$\frac{1}{3}y - \frac{1}{2} \leq \frac{5}{6}y + \frac{1}{2}$$
$$6\left(\frac{1}{3}y - \frac{1}{2}\right) \leq 6\left(\frac{5}{6}y + \frac{1}{2}\right)$$
$$2y - 3 \leq 5y + 3$$
$$-5y + 2y - 3 \leq -5y + 5y + 3$$
$$-3y - 3 + 3 \leq 3 + 3$$
$$-3y \leq 6$$
$$-\frac{1}{3}(-3y) \geq -\frac{1}{3}(6)$$
$$y \geq -2$$

See the graph in the back of the textbook.

45.
$$-2.8x + 8.4 < -14x - 2.8$$
$$-2.8x + 14x + 8.4 < 14x - 14x - 2.8$$
$$11.2x + 8.4 < -2.8$$
$$11.2x + 8.4 - 8.4 < -2.8 - 8.4$$
$$11.2x < -11.2$$
$$\frac{11.2x}{11.2} < \frac{-11.2}{11.2}$$
$$x < -1$$

See the graph in the back of the textbook.

47.
$$3(m - 2) - 4 \geq 7m + 14$$
$$3m - 6 - 4 \geq 7m + 14$$
$$3m - 10 \geq 7m + 14$$
$$-7m + 3m - 10 \geq -7m + 7m + 14$$
$$-4m - 10 \geq 14$$
$$-4m - 10 + 10 \geq 14 + 10$$
$$-4m \geq 24$$
$$-\frac{1}{4}(-4m) \leq -\frac{1}{4}(24)$$
$$m \leq -6$$

See the graph in the back of the textbook.

49.
$$3 - 4(x - 2) \leq -5x + 6$$
$$3 - 4x + 8 \leq -5x + 6$$
$$-4x + 11 \leq -5x + 6$$
$$-4x + 5x + 11 \leq -5x + 5x + 6$$
$$x + 11 \leq 6$$
$$x + 11 + (-11) \leq 6 + (-11)$$
$$x \leq -5$$

51.
$$3x + 2y < 6$$
$$2y < -3x + 6$$
$$y < -\frac{3}{2}x + 3$$

53.
$$2x - 5y > 10$$
$$-5y > -2x + 10$$
$$y < \frac{2}{5}x - 2$$

55.
$$-3x + 7y \leq 21$$
$$7y \leq 3x + 21$$
$$y \leq \frac{3}{7}x + 3$$

Problem Set 2.8

57. $2x - 4y \geq -4$
$-4y \geq -2x - 4$
$y \leq \dfrac{1}{2}x + 1$

59. $x < 3$

61. $x \geq 3$

63. Let $x =$ first integer
$x + 1 =$ next consecutive integer
$x + x + 1 \geq 583$
$2x + 1 \geq 583$
$2x \geq 582$
$x \geq 291$

65. Let x represent the number.
$2x + 6 < 10$
$2x < 4$
$x < 2$

67. Let x represent the number.
$4x > x - 8$
$3x > -8$
$x > -\dfrac{8}{3}$

69. Let $w =$ the width, $3w =$ the length. Using the formula for perimeter:
$2(w) + 2(3w) \geq 48$
$2w + 6w \geq 48$
$8w \geq 48$
$w \geq 6$

The width is at least 6 meters.

71. Let x, $x + 2$, and $x + 4 =$ the sides of the triangle.
$x + (x + 2) + (x + 4) > 24$
$3x + 6 > 24$
$3x > 18$
$x > 6$

The shortest side is an even number greater than 6 inches.

73. $t \geq 100$

75. Loss: $7.50x < 1500$
$x < 200$
Less than 200 tickets
Profit: $7.50x > 1500$
$x > 200$
Greater than 200 tickets

© 2000 Harcourt, Inc

77. b (commutative property of addition)

79. a (distributive property)

81. b and c (commutative and associative properties of addition)

2.9 Compound Inequalities

1-15. See the graph in the back of the textbook.

17. $3x - 1 < 5$ or $5x - 5 > 10$
    $3x < 6 \qquad\qquad 5x > 15$
    $x < 2 \qquad\qquad x > 3$

    See the graph in the back of the textbook.

19. $x - 2 > -5$ and $x + 7 < 13$
    $x > -3 \qquad\qquad x < 6$

    See the graph in the back of the textbook.

21. $11x < 22$ or $12x > 36$
    $x < 2 \qquad\qquad x > 3$

    See the graph in the back of the textbook.

23. $3x - 5 < 10$ and $2x + 1 > -5$
    $3x < 15 \qquad\qquad 2x > -6$
    $x < 5 \qquad\qquad x > -3$

    See the graph in the back of the textbook.

25. $2x - 3 < 8$ and $3x + 1 > -10$
    $2x < 11 \qquad\qquad 3x > -11$
    $x < \dfrac{11}{2} \qquad\qquad x > -\dfrac{11}{3}$

    See the graph in the back of the textbook.

27. $2x - 1 < 3$ and $3x - 2 > 1$
    $2x < 4 \qquad\qquad 3x > 3$
    $x < 2 \qquad\qquad x > 1$

    See the graph in the back of the textbook.

29. $-1 \leq x - 5 \leq 2$
    $4 \leq x \leq 7$

    See the graph in the back of the textbook.

31. $-4 \leq 2x \leq 6$
    $-2 \leq x \leq 3$

    See the graph in the back of the textbook.

33. $-3 < 2x + 1 < 5$
    $-4 < 2x < 4$
    $-2 < x < 2$

    See the graph in the back of the textbook.

35. $0 \leq 3x + 2 \leq 7$
    $-2 \leq 3x \leq 5$
    $-\dfrac{2}{3} \leq x \leq \dfrac{5}{3}$

    See the graph in the back of the textbook.

© 2000 Harcourt, Inc

Problem Set 2.9

37. $-7 < 2x+3 < 11$
$-10 < 2x < 8$
$-5 < x < 4$

See the graph in the back of the textbook.

39. $-1 \le 4x+5 \le 9$
$-6 \le 4x \le 4$
$-\dfrac{3}{2} \le x \le 1$

See the graph in the back of the textbook.

41. $-2 < x < 3$

43. $x \le -2$ or $x \ge 3$

45. (a) $2x+x > 10$; $x+10 > 2$; $2x+10 > x$

(b) $2x+x > 10$ $x+10 > 2x$
     $3x > 10$ $10 > x$
     $x > \dfrac{10}{3}$ $x < 10$
     $\dfrac{10}{3} < x < 10$

47. See the graph in the back of the textbook.

49. Let $x =$ the number
$10 < x+5 < 20$
$5 < x < 15$
The number is between 5 and 15.

51. Let $x =$ the number
$5 < 2x-3 < 7$
$8 < 2x < 10$
$4 < x < 5$
The number is between 4 and 5.

53. Let $w =$ the width, $w+4 =$ the length. Using the formula for perimeter:
$20 < 2w+2(w+4) < 30$
$20 < 2w+2w+8 < 30$
$20 < 4w+8 < 30$
$12 < 4w < 22$
$3 < w < \dfrac{11}{2}$

The width is between 3 inches and $\dfrac{11}{2} = 5\dfrac{1}{2}$ inches.

55. Simplifying the expression: $-|-5| = -(5) = -5$

57. Simplifying the expression: $-3 - 4(-2) = -3 + 8 = 5$

59. Simplifying the expression: $5|3-8| - 6|2-5| = 5|-5| - 6|-3| = 5(5) - 6(3) = 25 - 18 = 7$

61. Simplifying the expression: $5 - 2[-3(5-7) - 8] = 5 - 2[-3(-2) - 8] = 5 - 2(6-8) = 5 - 2(-2) = 5 + 4 = 9$

63. $-3 - (-9) = -3 + 9 = 6$

65. Applying the distributive property: $\dfrac{1}{2}(4x-6) = \dfrac{1}{2} \cdot 4x - \dfrac{1}{2} \cdot 6 = 2x - 3$

67. The integers are: $-3, 0, 2$

Chapter 2 Review

1. $5x - 8x = (5-8)x = -3x$
3. $-a + 2 + 5a - 9 = -a + 5a + 2 - 9 = (-1 + 5)a + (-7) = 4a - 7$
5. $6 - 2(3y + 1) - 4 = 6 - 6y - 2 - 4 = -6y$
7. $7x - 2$,   Letting $x = 3$ $7(3) - 2 = 19$
9. $-x - 2x - 3x = -6x$,   Letting $x = 3$ $-6(3) = -18$
11. $-3x + 2$,   Letting $x = -2$ $-3(-2) + 2 = 6 + 2 = 8$

13. $$\begin{aligned} x + 2 &= -6 \\ x + 2 + (-2) &= -6 + (-2) \\ x &= -8 \end{aligned}$$

15. $$\begin{aligned} 10 - 3y + 4y &= 12 \\ 10 + y &= 12 \\ -10 + 10 + y &= -10 + 12 \\ y &= 2 \end{aligned}$$

17. $$\begin{aligned} 2x &= -10 \\ \tfrac{1}{2}(2x) &= \tfrac{1}{2}(-10) \\ x &= -5 \end{aligned}$$

19. $$\begin{aligned} \tfrac{x}{3} &= 4 \\ 3\left(\tfrac{x}{3}\right) &= 3(4) \\ x &= 12 \end{aligned}$$

21. $$\begin{aligned} 3a - 2 &= 5a \\ -3a + 3a - 2 &= -3a + 5a \\ -2 &= 2a \\ \tfrac{1}{2}(-2) &= \tfrac{1}{2}(2a) \\ a &= -1 \end{aligned}$$

23. $$\begin{aligned} 3x + 2 &= 5x - 8 \\ 3x + (-5x) + 2 &= 5x + (-5x) - 8 \\ -2x + 2 &= -8 \\ -2x + 2 + (-2) &= -8 + (-2) \\ -2x &= -10 \\ -\tfrac{1}{2}(-2x) &= -\tfrac{1}{2}(-10) \\ x &= 5 \end{aligned}$$

25. $$\begin{aligned} 0.7x - 0.1 &= 0.5x - 0.1 \\ 0.7x + (-0.5x) - 0.1 &= 0.5x + (-0.5x) - 0.1 \\ 0.2x - 0.1 &= -0.1 \\ 0.2x - 0.1 + 0.1 &= -0.1 + 0.1 \\ 0.2x &= 0 \\ \tfrac{0.2x}{0.2} &= \tfrac{0}{0.2} \\ x &= 0 \end{aligned}$$

27. $$\begin{aligned} 2(x - 5) &= 10 \\ 2x - 10 &= 10 \\ 2x - 10 + 10 &= 10 + 10 \\ 2x &= 20 \\ \tfrac{1}{2}(2x) &= \tfrac{1}{2}(20) \\ x &= 10 \end{aligned}$$

29. $\dfrac{1}{2}(3t-2)+\dfrac{1}{2}=\dfrac{5}{2}$

$\dfrac{3}{2}t-1+\dfrac{1}{2}=\dfrac{5}{2}$

$\dfrac{3}{2}t-\dfrac{1}{2}=\dfrac{5}{2}$

$\dfrac{3}{2}t-\dfrac{1}{2}+\dfrac{1}{2}=\dfrac{5}{2}+\dfrac{1}{2}$

$\dfrac{3}{2}t=3$

$\dfrac{2}{3}\left(\dfrac{3}{2}t\right)=\dfrac{2}{3}(3)$

$t=2$

31. $2(3x+7)=4(5x-1)+18$

$6x+14=20x-4+18$

$6x+14=20x+14$

$6x+(-20x)+14=20x+(-20x)+14$

$-14x+14=14$

$-14x+14+(-14)=14+(-14)$

$-14x=0$

$-\dfrac{1}{14}(-14x)=-\dfrac{1}{14}(0)$

$x=0$

33. $4(5)-5y=20$

$20-5y=20$

$-5y=0$

$y=0$

35. $4(-5)-5y=20$

$-20-5y=20$

$-5y=40$

$y=-8$

37. $2x-5y=10$

$-5y=-2x+10$

$y=\dfrac{2}{5}x-2$

39. $\pi r^2 h = V$

$\dfrac{\pi r^2 h}{\pi r^2} = \dfrac{V}{\pi r^2}$

$h = \dfrac{V}{\pi r^2}$

41. $0.86(240)=x$

$x=206.4$

86% of 240 is 206.4

43. $2x+6=28$

$2x=22$

$x=11$

The number is 11.

45.

	Dollars Invested at 9%	Dollars Invested at 10%
Number of	x	$x+300$
Interest on	$0.09(x)$	$0.10(x+300)$

$0.09(x)+0.10(x+300)=125$

$0.09x+0.10x+30=125$

$0.19x+30=125$

$0.19x=95$

$x=500$

$x+300=800$

The man invested $500 at 9% and $800 at 10%.

47. $-2x < 4$
$-\frac{1}{2}(-2x) > -\frac{1}{2}(4)$
$x > -2$

49. $-\frac{a}{2} \leq -3$
$-2\left(-\frac{a}{2}\right) \geq -2(-3)$
$a \geq 6$

51. $-4x + 5 > 37$
$-4x > 32$
$x < -8$

See the graph in the back of the textbook.

53. $2(3t+1) + 6 \geq 5(2t+4)$
$6t + 2 + 6 \geq 10t + 20$
$6t + 8 \geq 10t + 20$
$-4t + 8 \geq 20$
$-4t \geq 12$
$t \leq -3$

See the graph in the back of the textbook.

55. $-5x \geq 25$ or $2x - 3 \geq 9$
$x \leq -5$ $2x \geq 12$
$x \leq$ $x \geq 6$

See the graph in the back of the textbook.

Cumulative Review: Chapters 1-2

1. $6 + 3(6+2) = 6 + 3(8) = 6 + 24 = 30$

3. $7 - 9 - 12 = 7 + (-9) + (-12) = 7 + (-21) = -14$

5. $\frac{1}{5}(10x) = \left(\frac{1}{5} \cdot 10\right)x = 2x$

7. $\left(-\frac{2}{3}\right)^3 = \left(-\frac{2}{3}\right)\left(-\frac{2}{3}\right)\left(-\frac{2}{3}\right) = -\frac{8}{27}$

9. $-\frac{3}{4} \div \frac{15}{16} = -\frac{3}{4} \cdot \frac{16}{15} = -\frac{48}{60} = -\frac{4}{5}$

11. $\frac{-4(-6)}{-9} = \frac{24}{-9} = -\frac{8}{3}$

13. $\frac{(5-3)^2}{5^2 - 3^2} = \frac{2^2}{25-9} = \frac{4}{16} = \frac{1}{4}$

15. $21 = 3 \cdot 7$
$35 = 5 \cdot 7$
$LCM = 3 \cdot 5 \cdot 7 = 105$
$\frac{4}{21} - \frac{9}{35} = \frac{4 \cdot 5}{21 \cdot 5} - \frac{9 \cdot 3}{35 \cdot 3} = \frac{20}{105} - \frac{27}{105} = -\frac{7}{105} = -\frac{7}{3 \cdot 5 \cdot 7} = -\frac{1}{3 \cdot 5} = -\frac{1}{15}$

© 2000 Harcourt, Inc

Cumulative Review: Chapters 1-2

17.
$$7x = 6x + 4$$
$$7x + (-6x) = 6x + (-6x) + 4$$
$$x = 4$$

19.
$$-\frac{3}{5}x = 30$$
$$-\frac{5}{3}\left(\frac{3}{5}x\right) = -\frac{5}{3}(30)$$
$$x = -50$$

21.
$$5x - 7 = x - 1$$
$$5x + (-x) - 7 = x + (-x) - 1$$
$$4x - 7 = -1$$
$$4x - 7 + 7 = -1 + 7$$
$$4x = 6$$
$$\frac{1}{4}(4x) = \frac{1}{4}(6)$$
$$x = \frac{3}{2}$$

23.
$$15 - 3(2t + 4) = 1$$
$$15 - 6t - 12 = 1$$
$$-6t + 3 = 1$$
$$-6t + 3 + (-3) = 1 + (-3)$$
$$-6t = -2$$
$$-\frac{1}{6}(-6t) = -\frac{1}{6}(-2)$$
$$t = \frac{1}{3}$$

25.
$$\frac{1}{3}(x - 6) = \frac{1}{4}(x + 8)$$
$$\frac{1}{3}x - 2 = \frac{1}{4}x + 2$$
$$12\left(\frac{1}{3}x - 2\right) = 12\left(\frac{1}{4}x + 2\right)$$
$$4x - 24 = 3x + 24$$
$$4x + (-3x) - 24 = 3x + (-3x) + 24$$
$$x - 24 = 24$$
$$x - 24 + 24 = 24 + 24$$
$$x = 48$$

27.
$$3x + 4y = 12$$
$$3x + (-3x) + 4y = 12 + (-3x)$$
$$4y = -3x + 12$$
$$\frac{1}{4}(4y) = \frac{1}{4}(-3x + 12)$$
$$y = -\frac{3}{4}x + 3$$

29.
$$-5x + 9 < -6$$
$$-5x + 9 + (-9) < 6 + (-9)$$
$$-5x < -15$$
$$-\frac{1}{5}(-5x) > -\frac{1}{5}(-15)$$
$$x > 3$$

See the graph in the back of the textbook

31. $-2 < x + 1 < 5$
$-3 < x < 4$

See the graph in the back of the textbook.

33. The opposite of $-\frac{2}{3}$ is $\frac{2}{3}$, the reciprocal is $-\frac{3}{2}$, and the absolute value is $\left|-\frac{2}{3}\right|=\frac{2}{3}$.

35. The pattern is to add -3, so the next term is: $-5+(-3)=-8$

37. $\frac{1}{4}(8x-4)=\frac{1}{4}\cdot 8x-\frac{1}{4}\cdot 4=2x-1$

39. $\frac{234}{312}=\frac{2\cdot 3\cdot 3\cdot 13}{2\cdot 2\cdot 2\cdot 3\cdot 13}=\frac{3}{2\cdot 2}=\frac{3}{4}$

41. Evaluating when $a=3$ and $b=-2$: $a^2-2ab+b^2=(3)^2-2(3)(-2)+(-2)^2=9+12+4=25$

43. Let $x =$ the number.
$$2x+7=31$$
$$2x=24$$
$$x=12$$

45. Let $x =$ the acute angle, and $90°$ is the right angle. The sum of the three angles is $180°$.
$$x+42°+90°=180°$$
$$x+132°=180°$$
$$x=48°$$
The other two angles are $48°$ and $90°$.

47. The other angle must be $90°-25°=65°$.

49.

	Dollars Invested at 5%	Dollars Invested at 6%
Number of	x	$x+200$
Interest on	$0.05(x)$	$0.06(x+200)$

$$0.05(x)+0.06(x+200)=56$$
$$0.05x+0.06x+12=56$$
$$0.11x+12=56$$
$$0.11x=44$$
$$x=400$$
$$x+200=600$$
You have $400 invested at 5% and $600 invested at 6%.

Chapter 2 Test

1. $3x + 2 - 7x + 3 = 3x - 7x + 2 + 3 = -4x + 5$
2. $4a - 5 - a + 1 = 4a - a - 5 + 1 = 3a - 4$
3. $7 - 3(y + 5) - 4 = 7 - 3y - 15 - 4 = 7 - 3y - 15 - 4 = -3y - 12$
4. $8(2x + 1) - 5(x - 4) = 16x + 8 - 5x + 20 = 16x - 5x + 8 + 20 = 11x + 28$
5. Evaluating when $x = -5$: $2x - 3 - 7x = -5x - 3 = -5(-5) - 3 = 25 - 3 = 22$
6. Evaluating when $x = 2$ and $y = 3$: $x^2 + 2xy + y^2 = (2)^2 + 2(2)(3) + (3)^2 = 4 + 12 + 9 = 25$

7. $\quad 2x - 5 = 7$
$\quad 2x - 5 + 5 = 7 + 5$
$\quad 2x = 12$
$\quad \frac{1}{2}(2x) = \frac{1}{2}(12)$
$\quad x = 6$

8. $\quad 2y + 4 = 5y$
$\quad -2y + 2y + 4 = -2y + 5y$
$\quad 4 = 3y$
$\quad \frac{1}{3}(4) = \frac{1}{3}(3y)$
$\quad y = \frac{4}{3}$

9. $\quad \frac{1}{2}x - \frac{1}{10} = \frac{1}{5}x + \frac{1}{2}$
$\quad 10\left(\frac{1}{2}x - \frac{1}{10}\right) = 10\left(\frac{1}{5}x + \frac{1}{2}\right)$
$\quad 5x - 1 = 2x + 5$
$\quad 5x + (-2x) - 1 = 2x + (-2x) + 5$
$\quad 3x - 1 = 5$
$\quad 3x - 1 + 1 = 5 + 1$
$\quad 3x = 6$
$\quad \frac{1}{3}(3x) = \frac{1}{3}(6)$
$\quad x = 2$

10. $\quad \frac{2}{5}(5x - 10) = -5$
$\quad 2x - 4 = -5$
$\quad 2x - 4 + 4 = -5 + 4$
$\quad 2x = -1$
$\quad \frac{1}{2}(2x) = \frac{1}{2}(-1)$
$\quad x = -\frac{1}{2}$

11. $\quad -5(2x + 1) - 6 = 19$
$\quad -10x - 5 - 6 = 19$
$\quad -10x - 11 = 19$
$\quad -10x - 11 + 11 = 19 + 11$
$\quad -10x = 30$
$\quad -\frac{1}{10}(-10x) = -\frac{1}{10}(30)$
$\quad x = -3$

12. $\quad 0.04x + 0.06(100 - x) = 4.6$
$\quad 0.04x + 6 - 0.06x = 4.6$
$\quad -0.02x + 6 = 4.6$
$\quad -0.02x + 6 + (-6) = 4.6 + (-6)$
$\quad -0.02x = -1.4$
$\quad \frac{-0.02x}{-0.02} = \frac{-1.4}{-0.02}$
$\quad x = 70$

© 2000 Harcourt, Inc

13. $2(t-4)+3(t+5)=2t-2$
    $2t-8+3t+15=2t-2$
    $5t+7=2t-2$
    $5t+(-2t)+7=2t+(-2t)-2$
    $3t+7=-2$
    $3t+7+(-7)=-2+(-7)$
    $3t=-9$
    $\frac{1}{3}(3t)=\frac{1}{3}(-9)$
    $t=-3$

14. $2x-4(5x+1)=3x+17$
    $2x-20x-4=3x+17$
    $-18x-4=3x+17$
    $-18x+(-3x)-4=3x+(-3x)+17$
    $-21x-4=17$
    $-21x-4+4=17+4$
    $-21x=21$
    $-\frac{1}{21}(21x)=-\frac{1}{21}(21)$
    $x=-1$

15. $0.15(38)=x$
    $5.7=x$
    15% of 38 is 5.7.

16. $0.12x=240$
    $\frac{0.12x}{0.12}=\frac{240}{0.12}$
    $x=2000$
    12% of 2,000 is 240.

17. Substituting $y=-2$:
    $2x-3(-2)=12$
    $2x+6=12$
    $2x=6$
    $x=3$

18. Substituting $V=88$, $\pi=\frac{22}{7}$, and $r=3$:
    $\frac{1}{3}\cdot\frac{22}{7}\cdot(3)^2 h=88$
    $\frac{66}{7}h=88$
    $\frac{7}{66}\left(\frac{66}{7}h\right)=\frac{7}{66}(88)$
    $h=\frac{28}{3}$ inches

19. Solving for y:
    $2x+5y=20$
    $5y=-2x+20$
    $y=-\frac{2}{5}x+4$

20. Solving for v:
    $x+vt+16t^2=h$
    $vt=h-x-16t^2$
    $v=\frac{h-x-16t^2}{t}$

Chapter 2 Test

21.

	Ten Years Ago	Now
Dave	$2x-10$	$2x$
Rick	$x-10$	x

$$2x-10+x-10 = 40$$
$$3x-20 = 40$$
$$3x = 60$$
$$x = 20$$
$$2x = 40$$

Rick is 20 and Fred is 40.

22. Let $w=$ the width and $2w=$ the length.
$$2(w)+2(2w) = 60$$
$$2w+4w = 60$$
$$6w = 60$$
$$w = 10$$
$$2w = 20$$

The width is 10 inches and the length is 20 inches.

23.

	Dimes	Quarters
Number	$x+7$	x
Value (cents)	$10(x+7)$	$25(x)$

$$10(x+7)+25(x) = 350$$
$$10x+70+25x = 350$$
$$35x+70 = 350$$
$$35x = 280$$
$$x = 8$$
$$x+7 = 15$$

He has 8 quarters and 15 dimes in his collection.

24.

	Dollars Invested at 7%	Dollars Invested at 9%
Number of	x	$x+600$
Interest on	$0.07(x)$	$0.09(x+600)$

$$0.07(x)+0.09(x+600) = 182$$
$$0.07x+0.09x+54 = 182$$
$$0.16x+54 = 182$$
$$0.16x = 128$$
$$x = 800$$
$$x+600 = 1400$$

She has $800 invested at 7% and $1,400 invested at 9%.

25.
$$2x+3 < 5$$
$$2x+3+(-3) < 5+(-3)$$
$$2x < 2$$
$$\frac{1}{2}(2x) < \frac{1}{2}(2)$$
$$x < 1$$

See the graph in the back of the textbook.

26.
$$-5a > 20$$
$$-\frac{1}{5}(-5a) < -\frac{1}{5}(20)$$
$$a < -4$$

See the graph in the back of the textbook.

27.
$$0.4-0.2x \geq 1$$
$$-0.4+0.4-0.2x \geq -0.4+1$$
$$-0.2x \geq 0.6$$
$$\frac{-0.2x}{-0.2} \leq \frac{0.6}{-0.2}$$
$$x \leq -3$$

See the graph in the back of the textbook.

28.
$$4-5(m+1) \leq 9$$
$$4-5m-5 \leq 9$$
$$-5m-1 \leq 9$$
$$-5m-1+1 \leq 9+1$$
$$-5m \leq 10$$
$$-\frac{1}{5}(-5m) \geq -\frac{1}{5}(10)$$
$$m \geq -2$$

See the graph in the back of the textbook.

© 2000 Harcourt, Inc

29. $3 - 4x \geq -5$ or $2x \geq 10$
       $-4x \geq -8$ $x \geq 5$
         $x \leq 2$ $x \geq 5$

     See the graph in the back of the textbook.

30. $-7 < 2x - 1 < 9$
     $-6 < 2x < 10$
     $-3 < x < 5$

     See the graph in the back of the textbook.

CHAPTER 3
Linear Equations and Inequalities in Two Variables
3.1 Paired Data and Graphing Ordered Pairs

1, 3, 5, 7, 9, 11, 13, 15, 17 - See the graph of these ordered pairs in the back of the textbook.

19. $(-4, 4)$

21. $(-4, 2)$

23. $(-3, 0)$

25. $(2, -2)$

27. $(-5, -5)$

29. Yes, see the graph in the back of the textbook.

31. No, see the graph in the back of the textbook.

33. Yes, see the graph in the back of the textbook.

35. No, see the graph in the back of the textbook.

37. Yes, see the graph in the back of the textbook.

39. No, see the graph in the back of the textbook.

41. No, see the graph in the back of the textbook.

43. No, see the graph in the back of the textbook.

45. Every point on this line has a y-coordinate of -3. See the graph in the back of the textbook.

47. They are on the y-axis.

49. Any three: $(0, 0), (5, 40), (10, 80), (15, 120), (20, 160), (25, 200), (30, 240), (35, 280), (40, 320)$

51. See the graph in the back of the textbook.

53. See the graph in the back of the textbook.

55. See the graph in the back of the textbook.

57. 3, 10, 17, 24, 31   Add seven to the previous number to produce the new number.

59. 3, 1, $\frac{1}{3}$, $\frac{1}{9}$, $\frac{1}{27}$ Multiply the previous by $\frac{1}{3}$ to produce the new number.

61. 7, 4, 1, -2, -5 Add -3 to the previous number to produce the new number.

63. 7, 21, 63, 189, 567   Multiply the previous by 3 to produce the new number.

65. 5, 6, 8, 11, 15   Add the next consecutive counting number to the previous number to produce the new number.

3.2 Solutions to Linear Equations in Two Variables

1. Substituting $x = 0, x = 3,$ and $y = -6$

   $2(0) + y = 6 \qquad 2(3) + y = 6 \qquad 2x + (-6) = 6$
   $0 + y = 6 \qquad\quad 6 + y = 6 \qquad\quad 2x = 12$
   $y = 6 \qquad\qquad y = 0 \qquad\qquad\quad x = 6$

   The ordered pairs are $(0, 6), (3, 0),$ and $(6, -6)$.

3. Substituting $x = 0, y = 0,$ and $x = -4$:

   $3(0) + 4y = 12 \qquad 3x + 4(0) = 12 \qquad 3(-4) + 4y = 12$
   $0 + 4y = 12 \qquad\quad 3x + 0 = 12 \qquad\quad -12 + 4y = 12$
   $4y = 12 \qquad\qquad 3x = 12 \qquad\qquad\quad 4y = 24$
   $y = 3 \qquad\qquad\quad x = 4 \qquad\qquad\qquad y = 6$

   The ordered pairs are $(0, 3), (4, 0),$ and $(-4, 6)$.

5. Substituting $x = 1, y = 0, x = 5$:

   $y = 4(1) - 3 \qquad 0 = 4x - 3 \qquad y = 4(5) - 3$
   $y = 4 - 3 \qquad\qquad 3 = 4x \qquad\qquad y = 20 - 3$
   $y = 1 \qquad\qquad\quad x = \dfrac{3}{4} \qquad\qquad y = 17$

   The ordered pairs are $(1,1), \left(\dfrac{3}{4}, 0\right),$ and $(5, 17)$.

7. Substituting $x = 2, y = 6, x = 0$:

   $y = 7(2) - 1 \qquad 6 = 7x - 1 \qquad y = 7(0) - 1$
   $y = 14 - 1 \qquad\quad 7 = 7x \qquad\qquad y = 0 - 1$
   $y = 13 \qquad\qquad\; x = 1 \qquad\qquad\;\; y = -1$

   The ordered pairs are $(2, 13), (1, 6),$ and $(0, -1)$.

9. Substituting $y = 4, y = -3,$ and $y = 0$ results (in each case) in $x = -5$. The ordered pairs are $(-5, 4), (-5, -3),$ and $(-5, 0)$.

11. See the table in the back of the textbook.
13. See the table in the back of the textbook.
15. See the table in the back of the textbook.
17. See the table in the back of the textbook.
19. See the table in the back of the textbook.

Problem Set 3.2

21. Substituting each ordered pair into the equation:
    $(2,3)$: $2(2)-5(3) = 4-15 = -11 \neq 10$
    $(0,-2)$: $2(0)-5(-2) = 0+10 = 10$
    $\left(\frac{5}{2},1\right)$: $2\left(\frac{5}{2}\right)-5(1) = 5-5 = 0 \neq 10$
    Only the ordered pair $(0,-2)$ is a solution.

23. Substituting each ordered pair into the equation:
    $(1,5)$: $7(1)-2 = 7-2 = 5$
    $(0,-2)$: $7(0)-2 = 0-2 = -2$
    $(-2,-16)$: $7(-2)-2 = -14-2 = -16$
    All the ordered pairs $(1,5)$, $(0,-2)$ and $(-2,-16)$ are solutions.

25. Substituting each ordered pair into the equation:
    $(1,6)$: $6(1) = 6$
    $(-2,-12)$: $6(-2) = -12$
    $(0,0)$: $6(0) = 0$
    All the ordered pairs $(1,6)$, $(-2,-12)$ and $(0,0)$ are solutions.

27. Substituting each ordered pair into the equation:
    $(1,1)$: $1+1 = 2 \neq 0$
    $(2,-2)$: $2+(-2) = 0$
    $(3,3)$: $3+3 = 6 \neq 0$
    Only the ordered pair $(2,-2)$ is a solution.

29. Since $x = 3$, the ordered pair $(5,3)$ cannot be a solution. The ordered pairs $(3,0)$ and $(3,-3)$ are solutions.

31. Substituting $w = 3$:
    $2l + 2(3) = 30$
    $2l + 6 = 30$
    $2l = 24$
    $l = 12$
    The length is 12 inches.

33. See the table and the graph in the back of the textbook.
35. See the table and the graph in the back of the textbook.

37. $y = 13 + 1.5x$ See the table and bar chart in the back of the textbook.

39. $y = 7 + 1.1x$ See the table and bar chart in the back of the textbook.

41. Substituting $x = 4$:
$$3(4) + 2y = 6$$
$$12 + 2y = 6$$
$$2y = -6$$
$$y = -3$$

43. Substituting $x = 0$: $y = -\frac{1}{3}(0) + 2 = 0 + 2 = 2$

45. Substituting $x = 2$: $y = \frac{3}{2}(2) - 3 = 3 - 3 = 0$

47. Solving for y:
$$5x + y = 4$$
$$y = -5x + 4$$

49. Solving for y:
$$3x - 2y = 6$$
$$-2y = -3x + 6$$
$$y = \frac{3}{2}x - 3$$

3.3 Graphing Linear Equations in Two Variables

1. The ordered pairs are (0, 4), (2, 2) and (4, 0). See the graph in the back of the textbook.

3. The ordered pairs are (0, 3), (2, 1), and (4, −1). See the graph in the back of the textbook.

5. The ordered pairs are (0, 0), (−2, −4) and (2, 4). See the graph in the back of the textbook.

7. The ordered pairs are (−3, −1), (0, 0), and (3, 1). See the graph in the back of the textbook.

9. The ordered pairs are (0, 1), (−1, −1) and (1, 3). See the graph in the back of the textbook.

11. The ordered pairs are (0, 4), (−1, 4) and (2, 4). See the graph in the back of the textbook.

13. The ordered pairs are (−2, 2), (0, 3) and (2, 4). See the graph in the back of the textbook.

15. The ordered pairs are (−3, 3), (0, 1) and (3, −1). See the graph in the back of the textbook.

17. Solving for y:
$$2x + y = 3$$
$$y = -2x + 3$$
The ordered pairs are $(-1, 5)$, $(0, 3)$, and $(1, 1)$.
See the graph in the back of the textbook.

19. Solving for y:
$$3x + 2y = 6$$
$$2y = -3x + 6$$
$$y = -\frac{3}{2}x + 3$$
The ordered pairs are $(0, 3)$, $(2, 0)$, and $(4, -3)$.
See the graph in the back of the textbook.

© 2000 Harcourt, Inc

Problem Set 3.3

21. Solving for y:
$$-x + 2y = 6$$
$$2y = x + 6$$
$$y = \frac{1}{2}x + 3$$
The ordered pairs are $(-2, 2)$, $(0, 3)$, and $(2, 4)$.
See the graph in the back of the textbook.

23. Three solutions are $(-4, 2)$, $(0, 0)$, and $(4, -2)$. See the graph in the back of the textbook.

25. Three solutions are $(-1, -4)$, $(0, -1)$ and $(1, 2)$. See the graph in the back of the textbook.

27. Solving for y:
$$-2x + y = 1$$
$$y = 2x + 1$$
Three are $(-2, -3)$, $(0, 1)$, and $(2, 5)$.
See the graph in the back of the textbook.

29. Solving for y:
$$3x + 4y = 8$$
$$4y = -3x + 8$$
$$y = -\frac{3}{4}x + 2$$
Three solutions are $(-4, 5)$, $(0, 2)$, and $(4, -1)$.
See the graph in the back of the textbook.

31. Three solutions are $(-2, -4)$, $(-2, 0)$, and $(-2, 4)$. See the graph in the back of the textbook.

33. Three solutions are $(-4, 2)$, $(0, 2)$, and $(4, 2)$. See the graph in the back of the textbook.

35. The ordered pairs are $(1, 4)$, $(2, 3)$, and $(3, 2)$. See the graph in the back of the textbook.

37. $y = x$: See the graph in the back of the textbook.

39. See the table in the back of the textbook.

41. The ordered pairs are $(-3, 3)$, $(-2, 2)$, $(-1, 1)$, $(0, 0)$, $(1, 1)$, $(2, 2)$ and $(3, 3)$. See the graph in the back of the textbook.

43. See the graph in the back of the textbook.

45. See the table and the graph in the back of the textbook.

47. See the table and the graph in the back of the textbook.

49. $3(x - 2) = 9$
$3x - 6 = 9$
$3x = 15$
$x = 5$

51. $2(3x - 1) + 4 = -10$
$6x - 2 + 4 = -10$
$6x + 2 = -10$
$6x = -12$
$x = -2$

© 2000 Harcourt, Inc

53. $6 - 2(4x - 7) = -4$
$6 - 8x + 14 = -4$
$-8x + 20 = -4$
$-8x = -24$
$x = 3$

55. $6\left(\dfrac{1}{2}x + 4\right) = 6\left(\dfrac{2}{3}x + 5\right)$
$3x + 24 = 4x + 30$
$-x + 24 = 30$
$-x = 6$
$x = -6$

3.4 More on Graphing: Intercepts

1. To find the x intercept, let $y = 0$:
$2x + 0 = 4$
$2x = 4$
$x = 2$
To find the y intercept, let $x = 0$:
$2(0) + y = 4$
$0 + y = 4$
$y = 4$
See the graph in the back of the textbook.

3. To find the x intercept, let $y = 0$:
$-x + 0 = 3$
$-x = 3$
$x = -3$
To find the y intercept, let $x = 0$:
$0 + y = 3$
$y = 3$
See the graph in the back of the textbook.

5. To find the x intercept, let $y = 0$:
$-x + 2(0) = 2$
$-x = 2$
$x = -2$
To find the y intercept, let $x = 0$:
$-0 + 2y = 2$
$2y = 2$
$y = 1$
See the graph in the back of the textbook.

7. To find the x intercept, let $y = 0$:
$5x + 2(0) = 10$
$5x = 10$
$x = 2$
To find the y intercept, let $x = 0$:
$5(0) + 2y = 10$
$2y = 10$
$y = 5$
See the graph in the back of the textbook.

9. To find the x intercept, let $y = 0$:
$4x - 2(0) = 8$
$4x = 8$
$x = 2$
To find the y intercept, let $x = 0$:
$4(0) - 2y = 8$
$-2y = 8$
$y = -4$
See the graph in the back of the textbook.

11. To find the x intercept, let $y = 0$:
$-4x + 5(0) = 20$
$-4x = 20$
$x = -5$
To find the y intercept, let $x = 0$:
$-4(0) + 5y = 20$
$5y = 20$
$y = 4$
See the graph in the back of the textbook.

© 2000 Harcourt, Inc

Problem Set 3.4

13. To find the x-intercept, let $y = 0$:
$$2x - 6 = 0$$
$$2x = 6$$
$$x = 3$$
To find the y-intercept, let $x = 0$:
$$y = 2(0) - 6$$
$$y = -6$$
See the graph in the back of the textbook.

15. To find the x-intercept, let $y = 0$:
$$2x + 2 = 0$$
$$2x = -2$$
$$x = -1$$
To find the y-intercept, let $x = 0$:
$$y = 2(0) + 2$$
$$y = 2$$
See the graph in the back of the textbook.

17. To find the x-intercept, let $y = 0$:
$$2x - 1 = 0$$
$$2x = 1$$
$$x = \frac{1}{2}$$
To find the y-intercept, let $x = 0$:
$$y = 2(0) - 1$$
$$y = -1$$
See the graph in the back of the textbook.

19. To find the x-intercept, let $y = 0$:
$$\frac{1}{2}x + 3 = 0$$
$$\frac{1}{2}x = -3$$
$$x = -6$$
To find the y-intercept, let $x = 0$:
$$y = \frac{1}{2}(0) + 3$$
$$y = 3$$
See the graph in the back of the textbook.

21. To find the x-intercept, let $y = 0$:
$$-\frac{1}{3}x - 2 = 0$$
$$-\frac{1}{3}x = 2$$
$$x = -6$$
To find the y-intercept, let $x = 0$:
$$y = -\frac{1}{3}(0) - 2$$
$$y = -2$$
See the graph in the back of the textbook.

23. Another point on the line is $(2, -4)$. See the graph in the back of the textbook.

25. Another point on the line is $(2, 4)$. See the graph in the back of the textbook.

27. Another point on the line is $(3, 1)$. See the graph in the back of the textbook.

29. Another point on the line is $(3, -1)$. See the graph in the back of the textbook.

31. Another point on the line is $(3, 2)$. See the graph in the back of the textbook.

33. The y-intercept is -4. See the graph in the back of the textbook.

35. The x-intercept is -3. See the graph in the back of the textbook.

37. The x and y-intercepts are both 3. See the graph in the back of the textbook.

© 2000 Harcourt, Inc

39. See the table in the back of the textbook.

41. The *x*-intercept is 3. See the graph in the back of the textbook.

43. The *y*-intercept is 4. See the graph in the back of the textbook.

45. To find the *x*-intercept, let $\theta = 0$:
$$\theta + 0 = 90$$
$$\alpha = 90$$
To find the θ-intercept, let $\alpha = 0$:
$$0 + \theta = 90$$
$$\theta = 90$$
See the graph in the back of the textbook.

47.
$$-3x \geq 12$$
$$-\frac{1}{3}(-3x) \leq -\frac{1}{3}(12)$$
$$x \leq -4$$

49.
$$-\frac{x}{3} \leq -1$$
$$-3\left(-\frac{x}{3}\right) \geq -3(-1)$$
$$x \geq 3$$

51.
$$-4x + 1 < 17$$
$$-4x + 1 - 1 < 17 - 1$$
$$-4x < 16$$
$$-\frac{1}{4}(-4x) > -\frac{1}{4}(16)$$
$$x > -4$$

3.5 The Slope of a Line

1. The slope is given by: $m = \dfrac{4-1}{4-2} = \dfrac{3}{2}$
   See the graph in the back of the textbook.

3. The slope is given by: $m = \dfrac{2-4}{5-1} = \dfrac{-2}{4} = -\dfrac{1}{2}$
   See the graph in the back of the textbook

5. The slope is given by: $m = \dfrac{2-(-3)}{4-1} = \dfrac{5}{3}$
   See the graph in the back of the textbook

7. The slope is given by: $m = \dfrac{3-(-2)}{1-(-3)} = \dfrac{5}{4}$
   See the graph in the back of the textbook

9. The slope is given by: $m = \dfrac{-2-2}{3-(-3)} = \dfrac{-4}{6} = -\dfrac{2}{3}$
   See the graph in the back of the textbook

11. The slope is given by: $m = \dfrac{-2-(-5)}{3-2} = \dfrac{3}{1} = 3$
    See the graph in the back of the textbook

13. $y = \dfrac{2}{3}x + 1$
    See the graph in the back of the textbook.

15. $y = \dfrac{3}{2}x - 3$
    See the graph in the back of the textbook

Problem Set 3.5

17. $y = -\dfrac{4}{3}x + 5$

    See the graph in the back of the textbook

19. $y = 2x + 1$

    See the graph in the back of the textbook1

21. $y = 3x - 1$

    See the graph in the back of the textbook

23. The y-intercept is 2, and the slope is given by:
$$m = \dfrac{5-(-1)}{1-(-1)} = \dfrac{6}{2} = 3$$

25. The y-intercept is -2, and the slope is given by:
$$m = \dfrac{2-0}{2-1} = \dfrac{2}{1} = 2$$

27. The slope is given by:
$$m = \dfrac{0-(-2)}{3-0} = \dfrac{2}{3}$$

    See the graph in the back of the textbook.

29. The slope is given by:
$$m = \dfrac{0-2}{4-0} = \dfrac{-2}{4} = -\dfrac{1}{2}$$

    See the graph in the back of the textbook.

31. The slope is 2 and the y-intercept is -3.

    See the graph in the back of the textbook.

33. The slope is $\dfrac{1}{2}$ and the y-intercept is 1.

    See the graph in the back of the textbook.

35. Using the slope formula:
$$\dfrac{y-2}{6-4} = 2$$
$$\dfrac{y-2}{2} = 2$$
$$y - 2 = 4$$
$$y = 6$$

37. $m_A = \dfrac{121-88}{1970-1960} = 3.3$

    $m_B = \dfrac{152-121}{1980-1970} = 3.1$

    $m_C = \dfrac{205-152}{1990-1980} = 5.3$

    $m_D = \dfrac{217-205}{2000-1990} = 1.2$

39. $m_A = \dfrac{28-33}{1999-1998} = -5$

    $m_B = \dfrac{23-25}{2001-2000} = -2$

    $m_C = \dfrac{20-22}{2003-2002} = -2$

41. When $x = -3$: $2x - 9 = 2(-3) - 9 = -6 - 9 = -15$

43. When $x = -3$: $9 - 6x = 9 - 6(-3) = 9 + 18 = 27$

45. When $x = -3$: $4(3x + 2) + 1 = 12x + 8 + 1 = 12x + 9 = 12(-3) + 9 = -36 + 9 = -27$

47. When $x = -3$: $2x^2 + 3x + 4 = 2(-3)^2 + 3(-3) + 4 = 18 - 9 + 4 = 13$

© 2000 Harcourt, Inc

3.6 Finding the Equation of a Line

1. The slope-intercept form is $y = \frac{2}{3}x + 1$.

3. The slope-intercept form is $y = \frac{3}{2}x - 1$.

5. The slope-intercept form is $y = -\frac{2}{5}x + 3$.

7. The slope-intercept form is $y = 2x - 4$.

9. The slope-intercept form is $y = -3x + 2$.

11. Solving for y:
    $$-2x + y = 4$$
    $$y = 2x + 4$$
    The slope is 2 and the y-intercept is 4.
    See the graph in the back of the textbook.

13. Solving for y:
    $$3x + y = 3$$
    $$y = -3x + 3$$
    The slope is -3 and the y-intercept is 3.
    See the graph in the back of the textbook

15. Solving for y:
    $$3x + 2y = 6$$
    $$2y = -3x + 6$$
    $$y = -\frac{3}{2}x + 3$$
    The slope is $-\frac{3}{2}$ and the y-intercept is 3.
    See the graph in the back of the textbook

17. Solving for y:
    $$4x - 5y = 20$$
    $$-5y = -4x + 20$$
    $$y = \frac{4}{5}x - 4$$
    The slope is $\frac{4}{5}$ and the y-intercept is -4.
    See the graph in the back of the textbook

19. Solving for y:
    $$-2x - 5y = 10$$
    $$-5y = 2x + 10$$
    $$y = -\frac{5}{2}x - 2$$
    The slope is $-\frac{2}{5}$ and the y-intercept is -2.
    See the graph in the back of the textbook

Problem Set 3.6

21. $y - y_1 = m(x - x_1)$
$y - (-5) = 2(x - (-2))$
$y + 5 = 2(x + 2)$
$y + 5 = 2x + 4$
$y = 2x - 1$

23. $y - y_1 = m(x - x_1)$
$y - 1 = -\dfrac{1}{2}(x - (-4))$
$y - 1 = -\dfrac{1}{2}(x + 4)$
$y - 1 = -\dfrac{1}{2}x - 2$
$y = -\dfrac{1}{2}x - 1$

25. $y - y_1 = m(x - x_1)$
$y - (-3) = \dfrac{3}{2}(x - 2)$
$y + 3 = \dfrac{3}{2}x - 3$
$y = \dfrac{3}{2}x - 6$

27. $y - y_1 = m(x - x_1)$
$y - 4 = -3(x - (-1))$
$y - 4 = -3(x + 1)$
$y - 4 = -3x - 3$
$y = -3x + 1$

29. $y - y_1 = m(x - x_1)$
$y - 4 = 1(x - 2)$
$y - 4 = x - 2$
$y = x + 2$

31. $m = \dfrac{-1 - (-4)}{1 - (-2)} = \dfrac{-1 + 4}{1 + 2} = \dfrac{3}{3} = 1$
$y - y_1 = m(x - x_1)$
$y - (-4) = 1(x - (-2))$
$y + 4 = x + 2$
$y = x - 2$

33. $m = \dfrac{1 - (-5)}{2 - (-1)} = \dfrac{1 + 5}{2 + 1} = \dfrac{6}{3} = 2$
$y - y_1 = m(x - x_1)$
$y - 1 = 2(x - 2)$
$y - 1 = 2x - 4$
$y = 2x - 3$

35. $m = \dfrac{6 - (-2)}{3 - (-3)} = \dfrac{6 + 2}{3 + 3} = \dfrac{8}{6} = \dfrac{4}{3}$
$y - y_1 = m(x - x_1)$
$y - 6 = \dfrac{4}{3}(x - 3)$
$y - 6 = \dfrac{4}{3}x - 4$
$y = \dfrac{4}{3}x + 2$

© 2000 Harcourt, Inc

37. $m = \dfrac{-5-(-1)}{3-(-3)} = \dfrac{-5+1}{3+3} = \dfrac{-4}{6} = -\dfrac{2}{3}$

$y - y_1 = m(x - x_1)$

$y - (-5) = -\dfrac{2}{3}(x - 3)$

$y + 5 = -\dfrac{2}{3}x + 2$

$y = -\dfrac{2}{3}x - 3$

39. $m = \dfrac{-2-1}{1-(-2)} = \dfrac{-2-1}{1+2} = \dfrac{-3}{3} = -1$

$y - y_1 = m(x - x_1)$

$y - (-2) = -1(x - 1)$

$y + 2 = -x + 1$

$y = -x - 1$

41. The y-intercept is 3 and the slope is: $m = \dfrac{3-0}{0-(1)} = \dfrac{3}{1} = 3$

$y = 3x + 3$

43. The y-intercept is -1 and the slope is: $m = \dfrac{0-(-1)}{4-0} = \dfrac{0+1}{4} = \dfrac{1}{4}$

$y = \dfrac{1}{4}x - 1$

45. The slope is $m = \dfrac{0-2}{3-0} = -\dfrac{2}{3}$

Since $b = 2$, the equation is $y = -\dfrac{2}{3}x + 2$.

47. The slope is $m = \dfrac{0-(-5)}{-2-0} = -\dfrac{5}{2}$

Since $b = -5$, the equation is $y = -\dfrac{5}{2}x - 5$.

49. The equation is $x = 3$.

51. The equation is $y = 3$.

53. (a) $75.00

(b) $300.00

(c) $m = \dfrac{0-300}{75-0} = -4$

(d) $y = -4x + 300$

55. $(0.25)(300) = x$

$x = 75$

75 is 25% of 300.

57. $0.15x = 60$

$x = \dfrac{60}{0.15}$

$x = 400$

60 is 15% of 400.

59. See the graph in the back of the textbook.

61. See the graph in the back of the textbook

63. See the graph in the back of the textbook.

3.7 Linear Inequalities in Two Variables

1. Checking the point $(0,0)$:
   $2(0)-3(0) = 0-0 < 6$ (true)
   See the graph in the back of the textbook.

3. Checking the point $(0,0)$:
   $0-2(0) = 0-0 \leq 4$ (true)
   See the graph in the back of the textbook.

5. Checking the point $(0,0)$:
   $0-0 \leq 2$ (true)
   See the graph in the back of the textbook.

7. Checking the point $(0,0)$:
   $3(0)-4(0) = 0-0 \geq 12$ (false)
   See the graph in the back of the textbook.

9. Checking the point $(0,0)$:
   $5(0)-0 = 0-0 \leq 5$ (true)
   See the graph in the back of the textbook.

11. Checking the point $(0,0)$:
    $2(0)+6(0) = 0+0 \leq 12$ (true)
    See the graph in the back of the textbook.

13. See the graph in the back of the textbook.

15. See the graph in the back of the textbook.

17. See the graph in the back of the textbook.

19. Checking the point $(0,0)$:
    $2(0)+0 = 0+0 > 3$ (false)
    See the graph in the back of the textbook.

21. Checking the point $(0,0)$:
    $0 \leq 3(0)-1$
    $0 \leq -1$ (false)
    See the graph in the back of the textbook.

23. Checking the point $(0,0)$:
    $0 \geq 0-5$
    $0 \geq -5$ (true)
    See the graph in the back of the textbook.

25. Checking the point $(0,0)$:
    $0 \leq -\dfrac{1}{2}(0)+2$
    $0 \leq 2$ (true)
    See the graph in the back of the textbook.

27. Checking the point $(0,0)$:
    $0 < -0+4$
    $0 < 4$ (true)
    See the graph in the back of the textbook.

29. $7-3(2x-4)-8 = 7-6x+12-8 = -6x+11$

31. $-\dfrac{3}{2}x = 12$
    $-\dfrac{2}{3}\left(-\dfrac{3}{2}x\right) = -\dfrac{2}{3}(12)$
    $x = -8$

© 2000 Harcourt, Inc

33. $8 - 2(x+7) = 2$
$8 - 2x - 14 = 2$
$-2x - 6 = 2$
$-2x = 8$
$x = -4$

35. $2x + 2w = P$
$2w = P - 2l$
$w = \dfrac{P - 2l}{2}$

37. $3 - 2x > 5$
$-3 + 3 - 2x > 5 - 3$
$-2x > 2$
$-\dfrac{1}{2}(-2x) < -\dfrac{1}{2}(2)$
$x < -1$

39. $3x - 2y \leq 12$
$3x - 3x - 2y \leq -3x + 12$
$-2y \leq -3x + 12$
$-\dfrac{1}{2}(-2y) \geq -\dfrac{1}{2}(-3x + 12)$
$y \geq \dfrac{3}{2}x - 6$

See the graph in the back of the textbook.

41. Let $w =$ the width and $3w + 5 =$ the length.
$2(w) + 2(3w + 5) = 26$
$2w + 6w + 10 = 26$
$8w + 10 = 26$
$w = 2$
$3w + 5 = 3(2) + 5 = 11$

The width is 2 inches and the length is 11 inches.

Chapter 3 Review

1. Substituting $x = 4$, $x = 0$, $y = 3$, and $y = 0$:

   $3(4) + y = 6$ $3(0) + y = 6$ $3x + 3 = 6$ $3x + 0 = 6$
   $12 + y = 6$ $0 + y = 6$ $3x = 3$ $3x = 6$
   $y = -6$ $y = 6$ $x = 1$ $x = 2$

   The ordered pairs are $(4, -6)$, $(0, 6)$, $(1, 3)$, and $(2, 0)$.

3. Substituting $x = 4$, $y = -2$, and $y = 3$:

   $y = 2(4) - 6$ $-2 = 2x - 6$ $3 = 2x - 6$
   $y = 8 - 6$ $4 = 2x$ $9 = 2x$
   $y = 2$ $x = 2$ $x = \dfrac{9}{2}$

   The ordered pairs are $(4, 2)$, $(2, -2)$, and $(\tfrac{9}{2}, 3)$.

© 2000 Harcourt, Inc

Chapter 3 Review

5. Substituting $x = 2$, $x = -1$, and $x = -3$ results (in each case) in $y = -3$. The ordered pairs are $(2, -3)$, $(-1, -3)$, and $(-3, -3)$.

7. Substituting each ordered pair into the equation:

$\left(-2, \dfrac{9}{2}\right)$: $3(-2) - 4\left(\dfrac{9}{2}\right) = -6 - 18 = -24 \neq 12$

$(0, 3)$: $\quad 3(0) - 4(3) = 0 - 12 = -12 \neq 12$

$\left(2, -\dfrac{3}{2}\right)$: $3(2) - 4\left(-\dfrac{3}{2}\right) = 6 + 6 = 12$

Only the ordered pair $\left(2, -\dfrac{3}{2}\right)$ is a solution.

9. See the graph in the back of the textbook.
11. See the graph in the back of the textbook.
13. See the graph in the back of the textbook.
15. The ordered pairs are $(-2, 0)$, $(0, -2)$, and $(1, -3)$.

See the graph in the back of the textbook.

17. The ordered pairs are $(1, 1)$, $(0, -1)$, and $(-1, -3)$.

See the graph in the back of the textbook.

19. See the graph in the back of the textbook.
21. See the graph in the back of the textbook.

23. To find the x-intercept, let $x = 0$:

$3x - 0 = 6$
$3x = 6$
$x = 2$

To find the y-intercept, let $x = 0$.
$3(0) - y = 6$
$-y = 6$
$y = -6$

25. To find the x-intercept, let $y = 0$:

$0 = x - 3$
$x = 3$

To find the y-intercept, let $x = 0$:
$y = 0 - 3$
$y = -3$

27. The slope is given by $m = \dfrac{5 - 3}{3 - 2} = \dfrac{2}{1} = 2$

29. The slope is given by $m = \dfrac{-8 - (-4)}{-3 - (-1)} = \dfrac{-8 + 4}{-3 + 1} = \dfrac{-4}{-2} = 2$

© 2000 Harcourt, Inc

31. Using the point-slope formula:
$$y - 4 = -2(x - (-1))$$
$$y - 4 = -2(x + 1)$$
$$y - 4 = -2x - 2$$
$$y = -2x + 2$$

33. Using the point-slope formula:
$$y - (-2) = -\frac{3}{4}(x - 3)$$
$$y + 2 = -\frac{3}{4}x + \frac{9}{4}$$
$$y = -\frac{3}{4}x + \frac{1}{4}$$

35. The slope-intercept form is $y = -x + 6$.

37. The equation is in slope-intercept form, so $m = 4$, $b = -1$.

39. Solving for y:
$$6x + 3y = 9$$
$$3y = -6x + 9$$
$$y = -2x + 3$$
$$m = -2 \text{ and } b = 3$$

41. Checking the point $(0, 0)$:
$0 - 0 < 3$ (true)
See the graph in the back of the textbook.

43. Checking the point $(0, 0)$:
$0 \le -2(0) + 3$
$0 \le 3$ (true)
See the graph in the back of the textbook.

Cumulative Review: Chapters 1-3

1. $7 - 2 \cdot 6 = 7 - 12 = -5$

3. $4 \cdot 6 + 12 \div 4 - 3^2 = 24 + 3 - 9 = 18$

5. $(4 - 9)(-3 - 8) = (-5)(-11) = 55$

7. $60 = 2 \cdot 2 \cdot 3 \cdot 5$
$84 = 2 \cdot 2 \cdot 3 \cdot 7$
$\text{LCM} = 2 \cdot 2 \cdot 3 \cdot 5 \cdot 7$
$\frac{11}{60} - \frac{13}{84} = \frac{11 \cdot 7}{60 \cdot 7} - \frac{13 \cdot 5}{84 \cdot 5} = \frac{77}{420} - \frac{65}{420} = \frac{12}{420} = \frac{2 \cdot 2 \cdot 3}{2 \cdot 2 \cdot 3 \cdot 5 \cdot 7} = \frac{1}{5 \cdot 7} = \frac{1}{35}$

9. $5a + 3 - 4a - 6 = 5a - 4a + 3 - 6 = a - 3$

11. $4x - 5 = 3$
$4x = 8$
$x = 2$

13. $6(t + 5) - 4 = 2$
$6t + 30 - 4 = 2$
$6t + 26 = 2$
$6t = -24$
$t = -4$

Cumulative Review: Chapters 1-3

15. $0.05x + 0.07(200 - x) = 11$
$0.05x + 14 - 0.07x = 11$
$-0.02x + 14 = 11$
$-0.02x = -3$
$x = 150$

17. $5 - 7x \geq 19$
$-5 + 5 - 7x \geq -5 + 19$
$-7x \geq 14$
$-\frac{1}{7}(-7x) \leq -\frac{1}{7}(14)$
$x \leq -2$
See the graph in the back of the textbook.

19. See the graph in the back of the textbook.

21. Checking the point $(0, 0)$:
$0 - 2(0) \leq 4$
$0 \leq 4$ (true)
See the graph in the back of the textbook.

23. The point (4, 7) does not lie on the line, while (1, 2) does lie on the line.
See the graph in the back of the textbook.

25. To find the *x*-intercept, let $y = 0$:
$2x + 5(0) = 10$
$2x = 10$
$x = 5$
To find the *y*-intercept, let $x = 0$:
$2(0) + 5y = 10$
$5y = 10$
$y = 2$
See the graph in the back of the textbook.

27. The slope is given by: $m = \frac{7-3}{5-2} = \frac{4}{3}$

29. Solving for *y*:
$2x + 3y = 6$
$3y = -2x + 6$
$y = -\frac{2}{3}x + 2$
The slope is $-\frac{2}{3}$.
See the graph in the back of the textbook.

31. $m = \frac{7-3}{4-2} = \frac{4}{2} = 2$
Using the point-slope formula:
$y - 3 = 2(x - 2)$
$y - 3 = 2x - 4$
$y = 2x - 1$

© 2000 Harcourt, Inc

33. $3x - 2(3) = 6 \qquad 3(0) - 2y = 6$
    $3x - 6 = 6 \qquad 0 - 2y = 6$
    $\ 3x = 12 \qquad{-2y = 6}$
    $\ x = 4 \qquad\ y = -3$

35. $x + 7 = 4$

    See the table in the back of the textbook.

37. $\left(\dfrac{1}{2}\right)^2 = \dfrac{1}{4}, \left(\dfrac{1}{3}\right)^2 = \dfrac{1}{9}, \left(\dfrac{1}{4}\right)^2 = \dfrac{1}{16},$ and $\left(\dfrac{1}{5}\right)^2 = \dfrac{1}{25}.$

    The next number in the sequence is $\left(\dfrac{1}{6}\right)^2 = \dfrac{1}{36}.$

39. $6(-2) - 5 = -12 - 5 = -17$

41. $\dfrac{75}{135} = \dfrac{3 \cdot 5 \cdot 5}{3 \cdot 3 \cdot 3 \cdot 5} = \dfrac{5}{3 \cdot 3} = \dfrac{5}{9}$

43. When $x = 2$: $x^2 + 6x - 7 = (2)^2 + 6(2) - 7 = 4 + 12 - 7 = 9$

45. $p \cdot 36 = 27$
    $p = \dfrac{27}{36}$
    $p = 0.75 = 75\%$
    75% of 36 is 27.

47. Let $w =$ the width and $2w + 5 =$ the length.
    $2(w) + 2(2w + 5) = 44$
    $2w + 4w + 10 = 44$
    $6w + 10 = 44$
    $6w = 34$
    $w = \dfrac{17}{3}$
    $2w + 5 = 2\left(\dfrac{17}{3}\right) + 5 = \dfrac{34}{3} + \dfrac{49}{3}$

    The width is $\dfrac{17}{3}$ cm and the length is $\dfrac{49}{3}$ cm.

49.

	Dollars Invested at 8%	Dollars Invested at 6%
Number of	$x + 900$	x
Interest on	$0.08(x+900)$	$0.06(x)$

$$0.08(x+900) + 0.06(x) = 8240$$
$$0.08x + 72 + 0.06x = 240$$
$$0.14x + 72 = 240$$
$$0.14x = 168$$
$$x = 1200$$
$$x + 900 = 2100$$

Barbara invested $1,200 at 6% and $2,100 at 8%.

Chapter 3 Test

1. Substituting $x = 0$, $y = 0$, $x = 10$, and $y = -3$:

   $2(0) - 5y = 10$ $2x - 5(0) = 10$ $2(10) - 5y = 10$ $2x - 5(-3) = 10$
   $0 - 5y = 10$ $2x - 0 = 10$ $20 - 5y = 10$ $2x + 15 = 10$
   $-5y = 10$ $2x = 10$ $-5y = -10$ $2x = -5$
   $y = -2$ $x = 5$ $y = 2$ $x = -\frac{5}{2}$

   The ordered pairs are $(0, -2)$, $(5, 0)$, $(10, 2)$ and $\left(-\frac{5}{2}, -3\right)$.

2. Substituting each ordered pair into the equation:

   $(2, 5)$: $4(2) - 3 = 8 - 3 = 5$
   $(0, -3)$: $4(0) - 3 = 0 - 3 = -3$
   $(3, 0)$: $4(3) - 3 = 12 - 3 = 9 \neq 0$
   $(-2, 11)$: $4(-2) - 3 = -8 - 3 = -11 \neq 11$

   The ordered pairs $(2, 5)$ and $(0, -3)$ are solutions.

3. See the graph in the back of the textbook.   4. See the graph in the back of the textbook.

5. To find the x-intercept, let $y = 0$:
$$3x - 5(0) = 15$$
$$3x = 15$$
$$x = 5$$
To find the y-intercept, let $x = 0$:
$$3(0) - 5y = 15$$
$$-5y = 15$$
$$y = -3$$

6. To find the x-intercept, let $y = 0$:
$$0 = \frac{3}{2}x + 1$$
$$-1 = \frac{3}{2}x$$
$$x = -\frac{2}{3}$$
To find the y-intercept, let $x = 0$:
$$y = \frac{3}{2}(0) + 1 = 1$$

7. The slope is given by: $m = \dfrac{-7-(-3)}{4-2} = \dfrac{-4}{2} = -2$

8. The slope is given by: $m = \dfrac{-8-5}{2-(-3)} = -\dfrac{13}{5}$

9. The slope is given by: $m = \dfrac{d-b}{c-a}$

10. The slope is given by: $m = \dfrac{4-3}{2x-5x} = -\dfrac{1}{3x}$

11. Using the point-slope formula:
$$y - 5 = 3(x - (-2))$$
$$y - 5 = 3(x + 2)$$
$$y - 5 = 3x + 6$$
$$y = 3x + 11$$

12. The slope-intercept form is $y = 4x + 8$.

13. $m = \dfrac{4-1}{-2-3} = -\dfrac{3}{5}$
Using the point-slope formula:
$$y - 1 = -\frac{3}{5}(x - 3)$$
$$y - 1 = -\frac{3}{5}x + \frac{9}{5}$$
$$y = -\frac{3}{5}x + \frac{14}{5}$$

14. $m = \dfrac{4-0}{3-1} = \dfrac{4}{2} = 2$
Using the point-slope formula:
$$y - 4 = 2(x - 3)$$
$$y - 4 = 2x - 6$$
$$y = 2x - 2$$

15. Checking the point $(0, 0)$:
$$0 < 0 + 4$$
$$0 < 4 \qquad \text{(true)}$$
See the graph in the back of the textbook.

16. Checking the point $(0, 0)$:
$$3(0) - 4(0) \geq 12$$
$$0 \geq 12 \qquad \text{(false)}$$
See the graph in the back of the textbook.

Chapter 4

Systems of Linear Equations

1. The intersection point is $(2,1)$.
   See the graph in the back of the textbook.

3. The intersection point is $(-1,2)$.
   See the graph in the back of the textbook.

5. The intersection point is $(3,5)$.
   See the graph in the back of the textbook.

7. The intersection point is $(4,3)$.
   See the graph in the back of the textbook.

9. The intersection point is $(0,-6)$.
   See the graph in the back of the textbook.

11. The intersection point is $(1,0)$.
    See the graph in the back of the textbook.

13. The intersection point is $(0,0)$.
    See the graph in the back of the textbook.

15. The intersection point is $(-5,-6)$.
    See the graph in the back of the textbook.

17. The intersection point is $(-1,-1)$.
    See the graph in the back of the textbook.

19. The intersection point is $(-3,2)$.
    See the graph in the back of the textbook.

21. The intersection point is $(-3,5)$.
    See the graph in the back of the textbook.

23. The intersection point is $(-4,6)$.
    See the graph in the back of the textbook.

25. There is no intersection (the line are parallel).
    See the graph in the back of the textbook.

27. The system is dependent (both lines are the same, they coincide). See the graph in the back of the textbook.

29. The intersection point is $\left(\frac{1}{2},1\right)$.
    See the graph in the back of the textbook.

31. The intersection point is $(2,1)$.
    See the graph in the back of the textbook.

33. (a) The lines intersect when the number of hours worked is 25.
    (b) For less than 20 hours the line for Gigi's is higher, so she should choose Gigi's.
    (c) For more than 30 hours, the line for Marcy's is higher, so she should choose Marcy's.

35. (a) The lines intersect when the number of hours used is 4.
    (b) For less than 3 hours, the line for Computer Services is lower, so she should choose Computer Services.
    (c) For more than 6 hours, the line for ICM is lower, so she should choose ICM.

37. Find the slope: $m = \dfrac{1-(-5)}{3-(-3)} = \dfrac{1+5}{3+3} = \dfrac{6}{6} = 1$

39. Find the slope: $m = \dfrac{0-3}{5-0} = -\dfrac{3}{5}$

41. See the graph in the back of the textbook.

© 2000 Harcourt, Inc

4.2 The Elimination Method

1. $x + y = 3$
   $\underline{x - y = 1}$
   $2x = 4$
   $x = 2$
   Substitute $x = 2$ into $x + y = 3$
   $2 + y = 3$
   $y = 1$
   The solution is $(2, 1)$.

3. $x + y = 10$
   $\underline{-x + y = 4}$
   $2y = 14$
   $y = 7$
   Substitute $y = 7$ into $x + y = 10$
   $x + y = 10 \quad x + 7 = 10$
   $x = 3$
   The solution is $(3, 7)$.

5. $x - y = 7$
   $\underline{-x - y = 3}$
   $-2y = 10$
   $y = -5$
   Substitute $y = -5$ into $x - y = 7$
   $x - (-5) = 7$
   $x + 5 = 7$
   $x = 2$
   The solution is $(2, -5)$.

7. $x + y = -1$
   $\underline{3x - y = -3}$
   $4x = -4$
   $x = -1$
   Substitute $x = -1$ into $x + y = -1$
   $-1 + y = -1$
   $y = 0$
   The solution is $(-1, 0)$.

9. $3x + 2y = 1$
   $\underline{-3x + -2y = -1}$
   $0 = 0$
   The lines coincide and there is an infinite number of solutions.

11. $3x - y = 4 \quad (1)$
    $2x + 2y = 24 \quad (2)$
    Multiply (1) by 2
    $6x - 2y = 8$
    $\underline{2x + 2y = 24}$
    $8x = 32$
    $x = 4$
    Substitute $x = 4$ into (2)
    $2(4) + 2y = 24$
    $8 + 2y = 24$
    $2y = 16$
    $y = 8$
    The solution is $(4, 8)$.

Problem Set 4.2

13. $5x - 3y = 2$ (1)
$10x - y = 1$ (2)
Multiply (2) by -3
$5x - 3y = -2$
$-30x + 3y = -3$
$\overline{-25x = -5}$
$x = \dfrac{1}{5}$
Substitute $x = \dfrac{1}{5}$ into (2)
$10\left(\dfrac{1}{5}\right) - y = 1$
$2 - y = 1$
$-y = -1$
$y = 1$
The solution is $\left(\dfrac{1}{5}, 1\right)$.

15. $11x - 4y = 11$ (1)
$5x + y = 5$ (2)
Multiply (2) by 4
$11x - 4y = 11$
$20x + 4y = 20$
$\overline{31x = 31}$
$x = 1$
Substitute $x = 1$ into (2)
$5(1) + y = 5$
$y = 0$
The solution is $(1, 0)$.

17. $3x - 5y = 7$ (1)
$-x + y = -1$ (2)
Multiply (2) by 3
$3x - 5y = 7$
$-3x + 3y = -3$
$\overline{-2y = 4}$
$y = -2$
Substitute $y = -2$ into (2)
$-x - 2 = -1$
$-x = 1$
$x = -1$
The solution is $(-1, -2)$.

19. $-x - 8y = -1$ (1)
$-2x + 4y = 13$ (2)
Multiply (1) by -2
$2x + 16y = 2$
$-2x + 4y = 13$
$\overline{20y = 15 \Rightarrow y = 3/4}$
Substitute $x = 3/4$ into (1)
$-x - 8(3/4) = -1$
$x = 5$
The solution is $(-5, 3/4)$.

© 2000 Harcourt, Inc

21. $-3x - y = 7$ (1)
    $6x + 7y = 11$ (2)
    Multiply (1) by 2
    $-6x - 2y = 14$
    $\underline{6x + 7y = 11}$
    $5y = 25$
    $y = 5$
    Substitute $y = 5$ into (1)
    $-3x - 5 = 7$
    $-3x = 12$
    $x = -4$
    The solution is $(-4, 5)$.

23. $6x - y = -8$
    $\underline{2x + y = -16}$
    $8x \quad = -24$
    $x \quad = -3$
    Substitute $x = -3$ into $2x + y = -16$
    $2(-3) + y = -16$
    $y = -10$
    The solution is $(-3, -10)$.

25. $x + 3y = 9$ (1)
    $2x - y = 4$ (2)
    Multiply (2) by 3
    $x + 3y = 9$
    $\underline{6x - 3y = 12}$
    $7x \quad = 21$
    $x = 3$
    Substitute $x = 3$ into (1)
    $3 + 3y = 9$
    $3y = 6$
    $y = 2$
    The solution is $(3, 2)$.

27. $x - 6y = 3$ (1)
    $4x + 3y = 21$ (2)
    Multiply (2) by 2
    $x - 6y = 3$
    $\underline{8x + 6y = 42}$
    $9x \quad = 45$
    $x = 5$
    Substitute $x = 5$ into (2)
    $4(5) + 3y = 21$
    $y = 1/3$
    The solution is $(5, 1/3)$.

29. $2x + 9y = 2$ (1)
    $5x + 3y = -8$ (2)
    Multiply (2) by -3
    $2x + 9y = 2$
    $\underline{-15x - 9y = 24}$
    $-13x \quad = 26$
    $x = -2$
    Substitute $x = -2$ into (1)
    $2(-2) + 9y = 2$
    $-4 + 9y = 2$
    $9y = 6$
    $y = \dfrac{2}{3}$
    The solution is $\left(-2, \dfrac{2}{3}\right)$.

31. $1/3x + 1/4y = 7/6$ (1)
    $3/2x - 1/3y = 7/3$ (2)
    Multiply (1) by 24 and (2) by 18
    $4x + 3y = 14$
    $\underline{27x - 6y = 42}$
    $35x \quad = 70 \Rightarrow x = 2$
    Substitute $x = 2$ into (2)
    $(3/2)2 - 1/3y = 7/3$
    $y = 2$
    The solution is $(2, 2)$.

Problem Set 4.2

33. $3x + 2y = -1$ (1)
$6x + 4y = 0$ (2)
Multiply (1) by -2
$-6x - 4y = 2$
$\underline{6x + 4y = 0}$
$0 = 2$
The lines are parallel, there is no solution.

35. $11x + 6y = 17$ (1)
$5x - 4y = 1$ (2)
Multiply (1) by 2 and (2) by 3
$22x + 12y = 34$
$\underline{15x - 12y = 3}$
$37x \quad\quad = 37$
$x = 1$
Substitute $x = 1$ into (1)
$11(1) + 6y = 17$
$11 + 6y = 17$
$6y = 6$
$y = 1$
The solution is $(1, 1)$.

37. $\frac{1}{2}x + \frac{1}{6}y = \frac{1}{3}$ (1)
$-x - \frac{1}{3}y = -\frac{1}{6}$ (2)
Multiply (1) by 12
Multiply (2) by 6
$6x + 2y = 4$
$\underline{-6x - 2y = -1}$
$0 = 3$
The lines are parallel, there is no solution.

39. $6x - 5y = 17$
$\underline{-3x + 5y = 4}$
$3x \quad\quad = 21$
$x \quad\quad = 7$
Substitute $x = 7$ into $-3x + 5y = 4$
$5y = 3(7) + 4 = 25$
$y = 5$
The solution is $(7, 5)$.

41. $x + y = 22$ (1)
$0.05x + 0.10y = 1.70$ (2)
Multiply (1) by -5
Multiply (2) by 100
$-5x - 5y = -110$
$\underline{5x + 10y = 170}$
$5y = 60$
$y = 12$
Substitute $y = 12$ into (1)
$x + 12 = 22$
$x = 10$
The solution is $(10, 12)$.

43. Solving for y:
$-2x + 4y = 8$
$4y = 2x + 8$
$y = \frac{1}{2}x + 2$
$m = \frac{1}{2}, b = 2$

45. Using the point-slope formula:
$$y-(-6)=3(x-(-2))$$
$$y+6=3(x+2)$$
$$y+6=3x+6$$
$$y=3x$$

47. $m = \dfrac{1-(-5)}{3-(-3)} = \dfrac{1+5}{3+3} = \dfrac{6}{6} = 1$

Using the point-slope formula:
$$y-1=1(x-3)$$
$$y-1=x-3$$
$$y=x-2$$

4.3 The Substitution Method

1. $x+y=11$ (1)
$y=2x-1$ (2)
Substitute (2) into (1)
$x+(2x-1)=11$
$3x-1=11$
$3x=12$
$x=4$
Substitute $x=4$ into (2)
$y=2(4)-1=8-1=7$
The solution is (4, 7).

3. $x+y=20$ (1)
$y=5x+2$ (2)
Substitute (2) into (1)
$x+(5x+2)=20$
$6x+2=20$
$6x=18$
$x=3$
Substitute $x=3$ into (2)
$y=5(3)+2=17$
The solution is $(3, 17)$.

5. $-2x+y=-1$ (1)
$y=-4x+8$ (2)
Substitute (2) into (1)
$-2x+(-4x+8)=-1$
$-6x+8=-1$
$x=\dfrac{-9}{-6}$
$x=\dfrac{3}{2}$
Substitute $x=\dfrac{3}{2}$ into (2)
$y=-4\left(\dfrac{3}{2}\right)+8=-6+8=2$
The solution is $\left(\dfrac{3}{2}, 2\right)$.

7. $3x-2y=-2$ (1)
$x=-y+6$ (2)
Substitute (2) into (1)
$3(-y+6)-2y=-2$
$-5y+18=-2$
$-5y=-20$
$y=4$
Substitute $y=4$ into (2)
$x=-4+6=2$
The solution is (2, 4).

Problem Set 4.3

9. $5x - 4y = -16 \quad (1)$
$y = 4 \quad (2)$
Substitute 4 for y in (1)
$5x - 4(4) = -16$
$5x - 16 = -16$
$x = 0$
$x = 0$
The solution is (0, 4).

11. $5x + 4y = 7 \quad (1)$
$y = -3x \quad (2)$
Substitute (2) into (1)
$5x + 4(-3x) = 7$
$-7x = 7$
$x = -1$
Substitute $x = -1$ into (2)
$y = -3(-1) = 3$
The solution is $(-1, 3)$.

13. $x + 3y = 4 \quad (1)$
$x - 2y = -1 \quad (2)$
Solve (2) for x and substitute into (1)
$x = 2y - 1 \quad (3)$
$(2y - 1) + 3y = 4$
$5y - 1 = 4$
$5y = 5$
$y = 1$
Substitute $y = 1$ into (3)
$x = 2(1) - 1 = 2 - 1 = 1$
The solution is (1, 1).

15. $2x + y = 1 \quad (1)$
$x - 5y = 17 \quad (2)$
Solve (2) for x and substitute into (1)
$x = 5y + 17 \quad (3)$
$2(5y + 17) + y = 1$
$11y + 34 = 1$
$11y = -33$
$y = -3$
Substitute $y = -3$ into (3)
$x = 5(-3) + 17 = 2$
The solution is $(2, -3)$.

17. $3x + 5y = -3 \quad (1)$
$x - 5y = -5 \quad (2)$
Solve (2) for x and substitute into (1)
$x = 5y - 5 \quad (3)$
$3(5y - 5) + 5y = -3$
$15y - 15 + 5y = -3$
$20y - 15 = -3$
$20y = 12$
$y = \dfrac{12}{20}$
$y = \dfrac{3}{5}$
Substitute $y = \dfrac{3}{5}$ into (3)
$x - 5\left(\dfrac{3}{5}\right) = -5$
$x - 3 = -5$
$x = -2$
The solution is $\left(-2, \dfrac{3}{5}\right)$.

19. $5x + 3y = 0 \quad (1)$
$x - 3y = -18 \quad (2)$
Solve (2) for x and substitute into (1)
$x = 3y - 18 \quad (3)$
$5(3y - 18) + 3y = 0$
$18y - 90 = 0$
$18y = 90$
$y = 5$
Substitute $y = 5$ into (3)
$x = 3(5) - 18 = -3$
The solution is $(-3, 5)$.

© 2000 Harcourt, Inc

21. $-3x - 9y = 7$ (1)
$x + 3y = 12$ (2)
Solve (2) for x and substitute into (1)
$-3(-3y + 12) - 9y = 7$
$9y - 36 - 9y = 7$
$-36 = 7$ False
The lines are parallel and there is no solution.

23. $5x - 8y = 7$ (1)
$y = 2x - 5$ (2)
Substitute (2) into (1)
$5x - 8(2x - 5) = 7$
$5x - 16x + 40 = 7$
$-11x + 40 = 7$
$-11x = -33$
$x = 3$
Substitute $x = 3$ into (2)
$y = 2(3) - 5 = 1$
The solution is $(3, 1)$.

25. $7x - 6y = -1$ (1)
$x = 2y - 1$ (2)
Substitute (2) into (1)
$7(2y - 1) - 6y = -1$
$14y - 7 - 6y = -1$
$8y - 7 = -1$
$8y = 6$
$y = \frac{6}{8} = \frac{3}{4}$
Substitute $y = \frac{3}{4}$ into (2)
$x = 2\left(\frac{3}{4}\right) - 1 = \frac{3}{2} - 1 = \frac{1}{2}$
The solution is $\left(\frac{1}{2}, \frac{3}{4}\right)$.

27. $-3x + 2y = 6$ (1)
$y = 3x$ (2)
Substitute (2) into (1)
$-3x + 2(3x) = 6$
$3x = 6$
$x = 2$
Substitute $x = 2$ into (2)
$y = 3(2) = 6$
The solution is $(2, 6)$.

29. $5x - 6y = -4$ (1)
$x = y$ (2)
Substitute (2) into (1)
$5y - 6y = -4$
$-y = -4$
$y = 4$
Substitute $y = 4$ into (2)
$x = 4$
The solution is $(4, 4)$.

31. $3x + 3y = 9$ (1)
$y = 2x - 12$ (2)
Substitute (2) into (1)
$3x + 3(2x - 12) = 9$
$3x + 6x - 36 = 9$
$9x - 36 = 9$
$9x = 45$
$x = 5$
Substitute $x = 5$ into (2)
$y = 2(5) - 12 = -2$
The solution is $(5, -2)$.

© 2000 Harcourt, Inc

Problem Set 4.3

33. $7x - 11y = 16$ (1)
          $y = 10$ (2)
 Substitute (2) into (1)
 $7x - 11(10) = 16$
 $7x - 110 = 16$
 $7x = 126$
 $x = 18$
 The solution is (18, 10).

35. $-4x + 4y = -8$ (1)
     $y = x - 2$ (2)
 Substitute (2) into (1)
 $-4x + 4(x - 2) = -8$
 $-8 = -8$
 A true statement. The graphs coincide and there is an infinite number of solutions.

37. $0.05x + 0.10y = 1.70$ (1)
         $y = 22 - x$ (2)
 Substitute (2) into (1)
 $0.05x + 0.10(22 - x) = 1.70$
 $0.05x + 2.2 - 0.10x = 1.70$
 $2.2 - 0.05x = 1.70$
 $-0.05x = -0.5$
 $x = 10$
 Substitute $x = 10$ into (2)
 $y = 22 - 10 = 12$
 The solution is (10, 12).

39. (a) The lines intersect at 1,000 miles.
 (b) For more than 1,200 miles, the line for the car is lower, so he should buy the car.
 (c) For fewer than 800 miles, the line for the truck is lower, so he should buy the truck.
 (d) Because the miles are ≥ 0.

41. Let x = number of 5-gallon bottles
 Let y = amount of the charge
 $y = 7 + (5x)(1.10) \rightarrow y = 7 + 5.5x$ (1)
 $y = 5 + (5x)(1.15) \rightarrow y = 5 + 5.75x$ (2)
 Substitute (2) into (1)
 $5 + 5.75x = 7 + 5.5x$
 $0.25x = 2$
 $x = 8$
 Substitute $x = 8$ into (1)
 $y = 7 + 5.50(8) = 51$
 Eight 5-gallon (total of 40 gallons) bottles must be used in a month for the two companies to charge the same amount of $51.

43. Let x = width, and $3x$ = length.
 $P = 2l + 2w$
 $24 = 2(3x) + 2x$
 $24 = 6x + 2x$
 $24 = 8x$
 $x = 3$ meters
 The width is 3 meters and length is 9 meters.

45. Let x = number of nickels
Let $3 + x$ = number of dimes
$$0.05x + 0.10(3 + x) = 2.10$$
$$0.15x = 1.80$$
$$x = 12$$
There are 12 nickels and 15 dimes.

47. What number is 8% of 6000?
$$N = (0.08) \cdot (6000)$$
$$N = 480$$

49. Let x = amount invested at 8%
Let $2x$ = amount invested at 10%
$$0.08x + 0.10(2x) = 224$$
$$0.08x + 0.20x = 224$$
$$0.28x = 224$$
$$x = 800$$
He invested \$800 at 8% and \$1600 at 10%.

4.4 Applications

1. Let x = first number
Let y = second number
$$x + y = 25 \quad (1)$$
$$y = 5 + x \quad (2)$$
Substitute (2) into (1)
$$x + (5 + x) = 25$$
$$2x + 5 = 25$$
$$2x = 20$$
$$x = 10$$
$$y = 5 + 10 = 15$$
The two numbers are 10 and 15.

3. Let x = first number
Let y = second number
$$x + y = 15 \quad (1)$$
$$x = 4y \quad (2)$$
Substitute (2) into (1)
$$4y + y = 15$$
$$5y = 15$$
$$y = 3$$
$$x = 4(3) = 12$$
The two numbers are 12 and 3.

5. Let x = larger number
Let y = smaller number
$$x - y = 5 \quad (1)$$
$$x = 2y + 1 \quad (2)$$
Substitute (2) into (1)
$$2y + 1 - y = 5$$
$$y + 1 = 5$$
$$y = 4$$
$$x = 2(4) + 1 = 9$$
The two numbers are 4 and 9.

7. Let x = first number
Let y = second number
$$x = 4y + 5 \quad (1)$$
$$x + y = 35 \quad (2)$$
Substitute (1) into (2)
$$4y + 5 + y = 35$$
$$5y = 30$$
$$y = 6$$
$$x = 4(6) + 5 = 29$$
The two numbers are 29 and 6.

Problem Set 4.4

9. Let x = amount invested at 6%
Let y = amount invested at 8%
$x + y = 20,000$ (1)
$0.06x + 0.08y = 1380$ (2)
Multiply (1) by -0.06
$-0.06x - 0.06y = -1200$
$\underline{0.06x + 0.08y = 1380}$
$0.02y = 180$
$y = 9000$
Substitute $y = 9000$ into (1)
$x + 9000 = 20,000$
$x = 11,000$
Mr. Wilson invested $9000 at 8% and 11,000 at 6%.

11. Let x = amount invested at 5%
Let y = amount invested at 6%
$x = 4y$ (1)
$0.05x + 0.06y = 520$ (2)
Substitute (1) into (2)
$0.05(4y) + 0.06y = 520$
$0.26y = 520$
$y = 2000$
$x = 4(2000) = 8000$
She invested $8000 at 5% and $2000 at 6%.

13. Let x = number of nickels
Let y = number of quarters
$x + y = 14$
$0.05x + 0.25y = 2.30$
Multiply (1) by -0.05
$-0.05x - 0.05y = -0.7$
$\underline{0.05x + 0.25y = 2.30}$
$0.20y = 1.6$
$y = 8$
Substitute $y = 8$ into (1)
$x + 8 = 14$
$x = 6$
Ron has 6 nickels and 8 quarters.

15. Let x = number of dimes
Let y = number of quarters
$x + y = 21$ (1)
$0.10x + 0.25y = 3.45$ (2)
Solve (1) for x
$x = 21 - y$ (3)
Substitute into (2)
$0.10(21 - y) + 0.25y = 3.45$
$2.1 - 0.10y + 0.25y = 3.45$
$0.15y = 1.35$
$y = 9$
Substitute $y = 9$ into (3)
$x = 21 - 9 = 12$
Tom has 12 dimes and 9 quarters.

17. Let x = number of liters of 50% alcohol solution
Let y = number of liters of 20% alcohol solution
$x + y = 18$ (1)
$0.50x + 0.20y = 0.30(18)$ (2)
Multiply (1) by -0.20
$-0.20x - 0.20y = -3.6$ (3)
$\underline{0.50x + 0.20y = 5.4}$
$0.30x = 1.8$
$x = 6$
Substitute $x = 6$ into (1)
$6 + y = 18$
$y = 12$
The mixture contains 6 liters of 50% alcohol solution and 12 liters of 20% alcohol solution.

19. Let x = amount of 10% disinfectant solution
Let y = amount of 7% disinfectant solution
$x + y = 30$ (1)
$0.10x + 0.07y = 30(0.08)$ (2)
Solve (1) for x
$x = 30 - y$ (3)
Substitute into (2)
$0.10(30 - y) + 0.07y = 2.4$
$3.00 - .10y + 0.07y = 2.4$
$-0.03y = -0.60$
$y = 20$
$x = 30 - 20 = 10$
10 gallons of 10% solution and 20 gallons of 7% solution should be used.

© 2000 Harcourt, Inc

21. Let x = number of adult tickets
Let y = number of kids tickets
$$x + y = 70 \quad (1)$$
$$5.50x + 4.00y = 310 \quad (2)$$
Mujltiply (1) by -4
$$-4.00x - 4.00y = -280$$
$$\underline{5.50x + 4.00y = 310}$$
$$1.5x = 30$$
$$x = 20$$
Substitute $x = 20$ into (1)
$$20 + y = 70$$
$$y = 50$$
The matinee had 20 adult tickets sold and 50 kids tickets sold.

23. Let x = width
Let y = length
$$2x + 2y = 96 \quad (1)$$
$$y = 2x \quad (2)$$
Substitute (2) into (1)
$$2x + 2(2x) = 96$$
$$6x = 96$$
$$x = 16$$
$$y = 2(16) = 32$$
The width is 16 feet and the length is 32 feet.

25. Let x = number of $5 chips
Let y = number of $25 chips
$$x + y = 45 \quad (1)$$
$$5x + 25y = 465 \quad (2)$$
Multiply (1) by -5
$$-5x - 5y = -225$$
$$\underline{-5x + 25y = 465}$$
$$20y = 240$$
$$y = 12$$
Substitute $y = 12$ into (1)
$$x + 12 = 45$$
$$x = 33$$
The gambler has 33 $5 chips and 12 $25 chips.

27. Let x = number of shares at $11 a share
Let y = number of shares at $20 a share
$$x + y = 150 \quad (1)$$
$$11x + 20y = 2550 \quad (2)$$
Solve (1) for x
$$x = 150 - y \quad (3)$$
Substitute into (2)
$$11(150 - y) + 20y = 2550$$
$$1650 - 11y + 20y = 2550$$
$$9y = 900$$
$$y = 100$$
$$x = 150 - 100 = 50$$
Mary Jo bought 50 shares at $11 a share and 100 shares at $20 a share.

29. Substituting $x = -2$, $x = 0$, and $x = 2$:
$$y = \frac{1}{2}(-2) + 3 = -1 + 3 = 2$$
$$y = \frac{1}{2}(0) + 3 = 0 + 3 = 3$$
$$y = \frac{1}{2}(2) + 3 = 1 + 3 = 4$$
The ordered pairs are $(-2, 2)$, $(0, 3)$, and $(2, 4)$.

31. See the graph in the back of the textbook.

33. $m = \dfrac{1-5}{0-2} = \dfrac{-4}{-2} = 2$

35. Using the point-slope formula:
$$y - 1 = \dfrac{1}{2}(x - (-2))$$
$$y - 1 = \dfrac{1}{2}(x + 2)$$
$$y - 1 = \dfrac{1}{2}x + 1$$
$$y = \dfrac{1}{2}x + 2$$

37. $m = \dfrac{1-5}{0-2} = \dfrac{-4}{-2} = 2$

Using the point-slope formula:
$$y - 1 = 2(x - 0)$$
$$y - 1 = 2x$$
$$y = 2x + 1$$

Chapter 4 Review

1. The intersection point is $(4, -2)$.
See the graph in the back of the textbook.

3. The intersection point is $(3, -2)$.
See the graph in the back of the textbook

5. The intersection point is $(2, 1)$.
See the graph in the back of the textbook

7. Adding the two equations:
$$2x = 2$$
$$x = 1$$
Substituting into the second equation:
$$1 + y = -2$$
$$y = -3$$
The solution is $(1, -3)$.

9. Multiplying the first equation by 2:
$$10x - 6y = 4$$
$$-10x + 6y = -4$$
Adding the two equations:
$$0 = 0$$
Since this statement is true, the system is dependent. The two lines coincide.

11. Multiplying the second equation by -4:
$$-3x + 4y = 1$$
$$16x - 4y = 12$$
Adding the two equations:
$$13x = 13$$
$$x = 1$$
Substituting into the second equation:
$$-4(1) + y = -3$$
$$-4 + y = -3$$
$$y = 1$$
The solution is $(1, 1)$.

13. Multiply the first equation by 3 and the second equation by 5:
$$-6x + 15y = -33$$
$$35x - 15y = -25$$
Adding the two equations:
$$29x = -58$$
$$x = -2$$
Substituting into the first equation:
$$-2(-2) + 5y = -11$$
$$4 + 5y = -11$$
$$5y = -15$$
$$y = -3$$
The solution is $(-2, -3)$.

15. Substituting into the first equation:
$$x + (-3x + 1) = 5$$
$$-2x + 1 = 5$$
$$-2x = 4$$
$$x = -2$$
Substituting into the second equation:
$$y = -3(-2) + 1 = 6 + 1 = 7$$

17. Substituting into the first equation:
$$4x - 3(3x + 7) = -16$$
$$4x - 9x - 21 = -16$$
$$-5x - 21 = -16$$
$$-5x = 5$$
$$x = -1$$
Substituting into the second equation:
$$y = 3(-1) + 7 = -3 + 7 = 4$$
The solution is $(-1, 4)$.

19. Solving the first equation for x:
$$x - 4y = 2$$
$$x = 4y + 2$$
Substituting into the second equation:
$$-3(4y + 2) + 12y = -8$$
$$-12y - 6 + 12y = -8$$
$$-6 = -8$$
Since this statement is false, there is no solution to the system. The two lines are parallel.

21. Solving the second equation for x:
$$x + 6y = -11$$
$$x = -6y - 11$$
Substituting into the second equation:
$$10(-6y - 11) - 5y = 20$$
$$-60y - 110 - 5y = 20$$
$$-65y - 110 = 20$$
$$-65y = 130$$
$$y = -2$$
Substituting into $x = -6y - 11$:
$$x = -6(-2) - 11 = 12 - 11 = 1$$
The solution is $(1, -2)$.

23. Let $x =$ smaller number
Let $y =$ larger number
$$x + y = 18$$
$$2x = 6 + y$$
Solving the first equation for y:
$$x + y = 18$$
$$y = -x + 18$$
Substituting into the second equation
$$2x = 6 + (-x + 18)$$
$$2x = -x + 24$$
$$3x = 24$$
$$x = 8$$
Substituting into the first equation:
$$8 + y = 18$$
$$y = 10$$
The two numbers are 8 and 10.

Chapter 4 Review

25. Let x = amount invested at 4%
Let y = amount invested at 5%
$$x + y = 12000$$
$$0.04x + 0.05y = 560$$
Multiplying the first equation by -0.04:
$$-0.04x - 0.04y = -480$$
$$0.04x + 0.05y = 560$$
Adding the two equations:
$$0.01y = 80$$
$$y = 8000$$
Substituting into the first equation:
$$x + 8000 = 12000$$
$$x = 4000$$
$4,000 was invested at 4% and $8,000 was invested at 5%.

27. Let x = number of dimes
Let y = number of nickels
$$x + y = 17$$
$$0.10x + 0.05y = 1.35$$
Multiplying the first equation by -0.05:
$$-0.05x - 0.05y = -0.85$$
$$0.10x + 0.05y = 1.35$$
Adding the two equations:
$$0.05x = 0.50$$
$$x = 10$$
Substituting into the first equation:
$$10 + y = 17$$
$$y = 7$$
Barbara has 10 dimes and 7 nickels.

29. Let x = liters of 20% alcohol solution
Let y = liters of 10% alcohol solution
$$x + y = 50$$
$$0.20x + 0.10y = 0.12(50)$$
Multiplying the first equation by -0.10:
$$-0.10x - 0.10y = -5$$
$$0.20x + 0.10y = 6$$
Adding the two equations:
$$0.10x = 1$$
$$x = 10$$
Substituting into the first equation:
$$10 + y = 50$$
$$y = 40$$
The solution contains 40 liters of 10% alcohol solution and 10 liters of 20% alcohol solution.

© 2000 Harcourt, Inc

Cumulative Review: Chapters 1-4

1. $3 \cdot 4 + 5 = 12 + 5 = 17$

3. $7[8 + (-5)] + 3(-7 + 12) = 7(3) + 3(5) = 21 + 15 = 36$

5. $8 - 6(5 - 9) = 8 - 6(-4) = 8 + 24 = 32$

7. $\dfrac{2}{3} + \dfrac{3}{4} - \dfrac{1}{6} = \dfrac{2 \cdot 4}{3 \cdot 4} + \dfrac{3 \cdot 3}{4 \cdot 3} - \dfrac{1 \cdot 2}{6 \cdot 2} = \dfrac{8}{12} + \dfrac{9}{12} - \dfrac{2}{12} = \dfrac{15}{12} = \dfrac{5}{4}$

9. $-5 - 6 = -y - 3 + 2y$
$-11 = y - 3$
$-11 + 3 = y - 3 + 3$
$y = -8$

11. $3(x - 4) = 9$
$3x - 12 = 9$
$3x - 12 + 12 = 9 + 12$
$3x = 21$
$\dfrac{1}{3}(3x) = \dfrac{1}{3}(21)$
$x = 7$

13. $0.3x + 0.7 \leq -2$
$0.3x + 0.7 - 0.7 \leq -2 - 0.7$
$0.3x \leq -2.7$
$\dfrac{0.3x}{0.3} \leq \dfrac{-2.7}{0.3}$
$x \leq -9$
See the graph in the back of the textbook.

15. See the graph in the back of the textbook.

17. See the graph in the back of the textbook.

19. The two lines are parallel.
See the graph in the back of the textbook.

21. Multiplying the first equation by -2:
$-2x - 2y = -14$
$2x + 2y = 14$
Adding the two equations:
$0 = 0$
Since this statement is true, the system is dependent. The two lines coincide.

23. Multiplying the second equation by 3:
$2x + 3y = 13$
$3x - 3y = -3$
Adding the two equations:
$5x = 10$
$x = 2$
Substituting into the first equation:
$2(2) + 3y = 13$
$4 + 3y = 13$
$y = 3$
The solution is $(2, 3)$.

© 2000 Harcourt, Inc

Cumulative Review: Chapters 1-4

25. Multiplying the second equation by -2:
$$2x + 5y = 33$$
$$-2x + 6y = 0$$
Adding the two equations:
$$11y = 33$$
$$y = 3$$
Substituting into the second equation:
$$x - 3(3) = 0$$
$$x - 9 = 0$$
$$x = 9$$
The solution is $(9, 3)$.

27. Multiplying the second equation by 7:
$$3x - 7y = 12$$
$$14x + 7y = 56$$
Adding the two equations:
$$17x = 68$$
$$x = 4$$
Substituting into the second equation:
$$2(4) + y = 8$$
$$8 + y = 8$$
$$y = 0$$
The solution is $(4, 0)$.

29. Substituting the value of y from the second equation into the first equation:
$$2x - 3(5x + 2) = 7$$
$$2x - 15x - 6 = 7$$
$$-13x - 6 = 7$$
$$-13x = 13$$
$$x = -1$$
Substituting into the second equation:
$$y = 5(-1) + 2 = -5 + 2 = -3$$
The solution is $(-1, -3)$.

31. Commutative property of addition

33. The quotient is: $\dfrac{-30}{6} = -5$

35. $p \cdot 82 = 20.5$
$$p = \frac{20.5}{82}$$
$$p = 0.25 = 25\%$$
25% of 82 is 20.5.

37. Simplifying, then evaluating when $x = 3$: $-3x + 7 + 5x = 2x + 7 = 2(3) + 7 = 6 + 7 = 13$

39. When $x = -2$: $4x - 5 = 4(-2) - 5 = -8 - 5 = -13$

41. $5 - (-8) = 5 + 8 = 13$

43. To find the x-intercept, let $y = 0$.
$$3x - 4(0) = 12$$
$$3x = 12$$
$$x = 4$$
To find the y-intercept, let $x = 0$:
$$3(0) - 4y = 12$$
$$-4y = 12$$
$$y = -3$$

45. $m = \dfrac{-4 - 1}{-5 - (-1)} = \dfrac{-5}{-4} = \dfrac{5}{4}$

47. The slope-intercept form is

$$y = \frac{2}{3}x + 3$$

49. $m = \dfrac{6-3}{6-4} = \dfrac{3}{2}$

Using the point-slope formula:

$$y - 3 = \frac{3}{2}(x - 4)$$
$$y - 3 = \frac{3}{2}x - 6$$
$$y = \frac{3}{2}x - 3$$

Chapter 4 Test

1. The intersection point is $(1, 2)$.
   See the graph in the back of the textbook.

2. The intersection point is $(3, -2)$.
   See the graph in the back of the textbook

3. Adding the two equations:
   $$3x = -9$$
   $$x = -3$$
   Substituting into the first equation:
   $$-3 - y = 1$$
   $$-y = 4$$
   $$y = -4$$
   The solution is $(-3, -4)$.

4. Multiplying the first equation by -1:
   $$-2x - y = -7$$
   $$3x + y = 12$$
   Adding the two equations:
   $$x = 5$$
   $$x = 1$$
   Substituting into the first equation:
   $$2(5) + y = 7$$
   $$10 + y = 7$$
   $$y = -3$$
   The solution is $(5, -3)$.

5. Multiplying the second equation by 4:
   $$7x + 8y = -2$$
   $$12x - 8y = 40$$
   Adding the two equations:
   $$19x = 38$$
   $$x = 2$$
   Substituting into the first equation:
   $$7(2) + 8y = -2$$
   $$14 + 8y = -2$$
   $$8y = -16$$
   $$y = -2$$
   The solution is $(2, -2)$.

6. Multiply the first equation by -3 and the second equation by 2:
   $$-18x + 30y = -18$$
   $$18x - 30y = 18$$
   Adding the two equations:
   $$0 = 0$$
   Since this equation is true, the system is dependent. The two lines coincide.

© 2000 Harcourt, Inc

Chapter 4 Test

7. Substituting the value of y from the second equation into the first equation:
$$3x + 2(2x+3) = 20$$
$$3x + 4x + 6 = 20$$
$$7x + 6 = 20$$
$$7x = 14$$
$$x = 2$$
Substituting into the second equation:
$$y = 2(2) + 3 = 4 + 3 = 7$$
The solution is $(2, 7)$.

8. Substituting x from the second equation into the first equation:
$$3(y+1) - 6y = -6$$
$$3y + 3 - 6y = -6$$
$$-3y + 3 = -6$$
$$-3y = -9$$
$$y = 3$$
Substituting into the second equation:
$$x = 3 + 1 = 4$$
The solution is $(4, 3)$.

9. Solving the second equation for y:
$$-3x + y = 3$$
$$y = 3x + 3$$
Substituting into the first equation:
$$7x - 2(3x+3) = -4$$
$$7x - 6x - 6 = -4$$
$$x - 6 = -4$$
$$x = 2$$
Substituting into $y = 3x + 3$:
$$y = 3(2) + 3 = 6 + 3 = 9$$
The solution is $(2, 9)$.

10. Solving the second equation for x:
$$x + 3y = -8$$
$$x = -3y - 8$$
Substituting into the first equation:
$$2(-3y-8) - 3y = -7$$
$$-6y - 16 - 3y = -7$$
$$-9y - 16 = -7$$
$$-9y = 9$$
$$y = -1$$
Substituting into $x = -3y - 8$:
$$x = -3(-1) - 8 = 3 - 8 = -5$$
The solution is $(-5, -1)$.

11. Let x and y represent the two numbers. The system of equation is:
$$x + y = 12$$
$$x - y = 2$$
Adding the two equations:
$$2x = 14$$
$$x = 7$$
Substituting into the first equation:
$$7 + y = 12$$
$$y = 5$$
The two numbers are 7 and 5.

12. Let x and y represent the two numbers. The system of equation is:
$$x + y = 15$$
$$y = 6 + 2x$$
Substituting into the first equation:
$$x + 6 + 2x = 15$$
$$3x + 6 = 15$$
$$3x = 9$$
$$x = 3$$
Substituting into the second equation:
$$y = 6 + 2(3) = 6 + 6 = 12$$
The two numbers are 3 and 12.

© 2000 Harcourt, Inc

13. Let x = amount invested at 9%
    Let y = amount invested at 11%
    $$x + y = 10000$$
    $$0.09x + 0.11y = 980$$
    Multiplying the first equation by -0.09:
    $$-0.09x - 0.09y = -900$$
    $$0.09x + 0.11y = 980$$
    Adding the two equations:
    $$0.02y = 80$$
    $$y = 4000$$
    Substituting into the first equation:
    $$x + 4000 = 10000$$
    $$x = 6000$$
    Dr. Stork should invest $6,000 at 9%.

14. Let x = number of nickels
    Let y = number of quarters
    $$x + y = 12$$
    $$0.05x + 0.25y = 1.60$$
    Multiplying the first equation by -0.05:
    $$-0.05x - 0.05y = -0.60$$
    $$0.05x + 0.25y = 1.60$$
    Adding the two equations:
    $$0.20y = 1.00$$
    $$y = 5$$
    Substituting into the first equation:
    $$x + 5 = 12$$
    $$x = 7$$
    Diane has 7 nickels and 5 quarters.

Chapter 5
Exponents and Polynomials

5.1 Multiplication with Exponents

1. The base is 4 and the exponent is 2. $4^2 = 4 \cdot 4 = 16$
3. The base is 0.3 and the exponent is 2. $(0.3)^2 = (0.3) \cdot (0.3) = 0.09$
5. The base is 4 and the exponent is 3. $4^3 = 4 \cdot 4 \cdot 4 = 64$
7. The base is -5 and the exponent is 2. $(-5)^2 = (-5) \cdot (-5) = 25$
9. The base is 2 and the exponent is 3. $-2^3 = -2 \cdot 2 \cdot 2 = -8$
11. The base is 3 and the exponent is 4. $3^4 = 3 \cdot 3 \cdot 3 \cdot 3 = 81$
13. The base is $\frac{2}{3}$ and the exponent is 2. $(\frac{2}{3})^2 = (\frac{2}{3}) \cdot (\frac{2}{3}) = \frac{4}{9}$
15. The base is $\frac{1}{2}$ and the exponent is 4. $(\frac{1}{2})^4 = (\frac{1}{2}) \cdot (\frac{1}{2}) \cdot (\frac{1}{2}) \cdot (\frac{1}{2}) = \frac{1}{16}$
17. (a) See the table in the back of the textbook. (b) For numbers larger than 1, the square of the number is greater than the number.
19. $x^4 \cdot x^5 = x^{4+5} = x^9$
21. $y^{10} \cdot y^{20} = y^{10+20} = y^{30}$
23. $2^5 \cdot 2^4 \cdot 2^3 = 2^{5+4+3} = 2^{12}$
25. $x^4 \cdot x^6 \cdot x^8 \cdot x^{10} = x^{4+6+8+10} = x^{28}$
27. $(x^2)^5 = x^{2 \cdot 5} = x^{10}$
29. $(5^4)^3 = 5^{4 \cdot 3} = 5^{12}$
31. $(y^3)^3 = y^{3 \cdot 3} = y^9$
33. $(2^5)^{10} = 2^{5 \cdot 10} = 2^{50}$
35. $(a^3)^x = a^{3x}$
37. $(b^x)^y = b^{xy}$
39. $(4x)^2 = 4^2 \cdot x^2 = 16x^2$
41. $(2y)^5 = 2^5 \cdot y^5 = 32y^5$
43. $(-3x)^4 = (-3)^4 \cdot x^4 = 81x^4$
45. $(0.5ab)^2 = (0.5)^2 \cdot a^2 b^2 = 0.25 a^2 b^2$
47. $(4xyz)^3 = 4^3 \cdot x^3 y^3 z^3 = 64 x^3 y^3 z^3$
49. $(2x^4)^3 = 2^3 (x^4)^3 = 8x^{12}$
51. $(4a^3)^2 = 4^2(a^3)^2 = 16a^6$
53. $(x^2)^3(x^4)^2 = x^6 \cdot x^8 = x^{14}$
55. $(a^3)^1 (a^2)^4 = a^3 \cdot a^8 = a^{11}$
57. $(2x)^3 (2x)^4 = (2x)^7 = 2^7 x^7 = 128 x^7$
59. $(3x^2)^3 (2x)^4 = 3^3 x^6 \cdot 2^4 x^4 = 27x^6 \cdot 16x^4 = 432 x^{10}$
61. $(4x^2 y^3)^2 = 4^2 x^4 y^6 = 16 x^4 y^6$
63. $(\frac{2}{3} a^4 b^5)^3 = (\frac{2}{3})^3 a^{12} b^{15} = \frac{8}{27} a^{12} b^{15}$
65. See the table and the graph in the back of the textbook.
67. See the table in the back of the textbook. They will create a smoother curve.
69. $43,200 = 4.32 \times 10^4$
71. $570 = 5.7 \times 10^2$
73. $238,000 = 2.38 \times 10^5$
75. $2.49 \times 10^3 = 2,490$
77. $3.52 \times 10^2 = 352$
79. $2.8 \times 10^4 = 28,000$
81. $V = (3 \text{ in.})^3 = 27 \text{ inches}^3$
83. $V = (2.5 \text{ in.})^3 = 15.6 \text{ inches}^3$
85. $V = (8 \text{ in.})(4.5 \text{ in.})(1 \text{ in.}) = 36 \text{ inches}^3$
87. Yes. If the box was 7 ft $\times$ 3 ft $\times$ 2 ft you could fit inside of it.
89. $650,000,000$ seconds $= 6.5 \times 10^8$ seconds
91. 7.4×10^5 dollars $= \$740,000$
93. 1.8×10^5 dollars $= \$180,000$

Use $d = \pi \cdot s \cdot c \cdot (\frac{1}{2} \cdot b)^2$ in Exercises 95, and 97.

95. $d = (3.14)(3.11)(8)(\frac{3.35}{2})^2 = 219 \text{ inches}^3$
97. $d = (3.14)(2.99)(6)(\frac{3.59}{2})^2 = 182 \text{ inches}^3$

99. (a) Stage 1: $\frac{4}{3} L_0 = (\frac{4}{3})(1 \text{ foot}) = \frac{4}{3}$ feet (b) Stage 2: $(\frac{4}{3})^2 L_0 = \frac{16}{9}$ feet
   (c) See the drawing in the back of the textbook. Stage 3: $(\frac{4}{3})^{10} L_0 = (\frac{4}{3})^3 = \frac{64}{27}$ feet
   (d) Stage 10: $(\frac{4}{3})^{10} L_0 = (\frac{4}{3})^{10}$ feet

101. $4 - 7 = 4 + (-7) = -3$
103. $4 - (-7) = 4 + 7 = 11$
105. $15 - 20 = 15 + (-20) = -5$
107. $-15 - (-20) = -15 + 20 = 5$

© 2000 Harcourt, Inc

5.2 Division with Exponents

1. $3^{-2} = \frac{1}{3^2} = \frac{1}{9}$

3. $6^{-2} = \frac{1}{6^2} = \frac{1}{36}$

5. $8^{-2} = \frac{1}{8^2} = \frac{1}{64}$

7. $5^{-3} = \frac{1}{5^3} = \frac{1}{125}$

9. $2x^{-3} = 2 \cdot \frac{1}{x^3} = \frac{2}{x^3}$

11. $(2x)^{-3} = \frac{1}{(2x)^3} = \frac{1}{8x^3}$

13. $(5y)^{-2} = \frac{1}{(5y)^2} = \frac{1}{25y^2}$

15. $10^{-2} = \frac{1}{10^2} = \frac{1}{100}$

17. See the table in the back of the textbook.

19. $\frac{5^1}{5^3} = 5^{1-3} = 5^{-2} = \frac{1}{5^2} = \frac{1}{25}$

21. $\frac{x^{10}}{x^4} = x^{10-4} = x^6$

23. $\frac{4^3}{4^0} = 4^{3-0} = 4^3 = 64$

25. $\frac{(2x)^7}{(2x)^4} = (2x)^{7-4} = (2x)^3 = 2^3 x^3 = 8x^3$

27. $\frac{6^{11}}{6} = \frac{6^{11}}{6^1} = 6^{11-1} = 6^{10} = 60,466,176$

29. $\frac{6}{6^{11}} = \frac{6^1}{6^{11}} = 6^{1-11} = 6^{-10} = \frac{1}{6^{10}} = \frac{1}{60,466,176}$

31. $\frac{2^{-5}}{2^3} = 2^{-5-3} = 2^{-8} = \frac{1}{2^8} = \frac{1}{256}$

33. $\frac{2^5}{2^{-3}} = 2^{5-(-3)} = 2^{5+3} = 2^8 = 256$

35. $\frac{(3x)^{-5}}{(3x)^{-8}} = (3x)^{-5-(-8)} = (3x)^{-5+8} = (3x)^3 = 3^3 x^3 = 27x^3$

37. $(3xy)^4 = 3^4 x^4 y^4 = 81 x^4 y^4$

39. $10^0 = 1$

41. $(2a^2 b)^1 = 2a^2 b$

43. $(7y^3)^{-2} = \frac{1}{(7y^3)^2} = \frac{1}{49 y^6}$

45. $x^{-3} \cdot x^{-5} = x^{-3-5} = x^{-8} = \frac{1}{x^8}$

47. $y^7 \cdot y^{-10} = y^{7-10} = y^{-3} = \frac{1}{y^3}$

49. $\frac{(x^2)^3}{x^4} = \frac{x^6}{x^4} = x^{6-4} = x^2$

51. $\frac{(a^4)^3}{(a^3)^2} = \frac{a^{12}}{a^6} = a^{12-6} = a^6$

53. $\frac{y^7}{(y^2)^8} = \frac{y^7}{y^{16}} = y^{7-16} = y^{-9} = \frac{1}{y^9}$

55. $\left(\frac{y^7}{y^2}\right)^8 = (y^{7-2})^8 = (y^5)^8 = y^{40}$

Problem Set 5.2

57. $\dfrac{(x^{-2})^3}{x^{-5}} = \dfrac{x^{-6}}{x^{-5}} = x^{-6-(-5)} = x^{-6+5} = x^{-1} = \dfrac{1}{x}$

59. $\left(\dfrac{x^{-2}}{x^{-5}}\right)^3 = (x^{-2+5})^3 = (x^3)^3 = x^9$

61. $\dfrac{(a^3)^2 (a^4)^5}{(a^5)^2} = \dfrac{a^6 \cdot a^{20}}{a^{10}} = \dfrac{a^{26}}{a^{10}} = a^{26-10} = a^{16}$

63. $\dfrac{(a^{-2})^3 (a^4)^2}{(a^{-3})^{-2}} = \dfrac{a^{-6} \cdot a^8}{a^6} = \dfrac{a^2}{a^6} = a^{2-6} = a^{-4} = \dfrac{1}{a^4}$

65. See the table and the graph in the back of the textbook.

67. $0.0048 = 4.8 \times 10^{-3}$

69. $25 = 2.5 \times 10^1$

71. $0.000009 = 9 \times 10^{-6}$

73. See the table in the back of the textbook.

75. $4.23 \cdot 10^{-3} = 0.00423$

77. $8 \times 10^{-5} = 0.00008$

79. $4.2 \times 10^0 = 4.2$

81. 2×10^{-3} seconds $= 0.002$ seconds

83. 0.006 inches $= 6 \times 10^{-3}$ inches

85. $25 \times 10^3 = 2.5 \times 10^4$

87. $23.5 \times 10^4 = 2.35 \times 10^5$

89. $0.82 \times 10^{-3} = 8.2 \times 10^{-4}$

91. The area of the smaller square is $(10 \text{ in.})^2 = 100$ inches^2, while the area of the larger square is $(20 \text{ in.})^2 = 400$ inches^2. It would take 4 smaller squares to cover the larger square.

93. The area of the smaller square is x^2, while the area of the larger square is $(2x)^2 = 4x^2$. It would take 4 smaller squares to cover the larger square.

95. The volume of the smaller box is $(6 \text{ in.})^3 = 216$ inches^3, while the volume of the larger box is $(12 \text{ in.})^3 = 1,728$ inches^3. Thus 8 smaller boxes will fit inside the larger box $(8 \cdot 216 = 1,728)$.

97. The volume of the smaller box is x^3, while the volume of the larger box is $(2x)^3 = 8x^3$. Thus 8 smaller boxes will fit inside the larger box.

99. $4x + 3x = (4+3)x = 7x$

101. $5a - 3a = (5-3)a = 2a$

103. $4y + 5y + y = (4+5+1)y = 10y$

© 2000 Harcourt, Inc

5.3 Operations with Monomials

1. $(3x^4)(4x^3) = 12x^{4+3} = 12x^7$

3. $(-2y^4)(8y^7) = -16y^{4+7} = -16y^{11}$

5. $(8x)(4x) = 32x^{1+1} = 32x^2$

7. $(10a^3)(10a)(2a^2) = 200a^{3+1+2} = 200a^6$

9. $(6ab^2)(-4a^2b) = -24a^{1+2}b^{2+1} = -24a^3b^3$

11. $(4x^2y)(3x^3y^3)(2xy^4) = 24x^{2+3+1}y^{1+3+4} = 24x^6y^8$

13. $\dfrac{15x^3}{5x^2} = \dfrac{15}{5} \cdot \dfrac{x^3}{x^2} = 3x$

15. $\dfrac{18y^9}{3y^{12}} = \dfrac{18}{3} \cdot \dfrac{y^9}{y^{12}} = 6 \cdot \dfrac{1}{y^3} = \dfrac{6}{y^3}$

17. $\dfrac{32a^3}{64a^4} = \dfrac{32}{64} \cdot \dfrac{a^3}{a^4} = \dfrac{1}{2} \cdot \dfrac{1}{a} = \dfrac{1}{2a}$

19. $\dfrac{21a^2b^3}{-7ab^5} = \dfrac{21}{-7} \cdot \dfrac{a^2}{a} \cdot \dfrac{b^3}{b^5} = -3 \cdot a \cdot \dfrac{1}{b^2} = -\dfrac{3a}{b^2}$

21. $\dfrac{3x^3y^2z}{27xy^2z^3} = \dfrac{3}{27} \cdot \dfrac{x^3}{x} \cdot \dfrac{y^2}{y^2} \cdot \dfrac{z}{z^3} = \dfrac{1}{9} \cdot x^2 \cdot \dfrac{1}{z^2} = \dfrac{x^2}{9z^2}$

23. See the table in the back of the textbook.

25. $(3 \times 10^3)(2 \times 10^5) = 6 \times 10^8$

27. $(3.5 \times 10^4)(5 \times 10^{-6}) = 17.5 \times 10^{-2} = 1.75 \times 10^{-1}$

29. $(5.5 \times 10^{-3})(2.2 \times 10^{-4}) = 12.1 \times 10^{-7} = 1.21 \times 10^{-6}$

31. $\dfrac{8.4 \times 10^5}{2 \times 10^2} = 4.2 \times 10^3$

33. $\dfrac{6 \times 10^8}{2 \times 10^{-2}} = 3 \times 10^{10}$

35. $\dfrac{2.5 \times 10^{-6}}{5 \times 10^{-4}} = 0.5 \times 10^{-2} = 5.0 \times 10^{-3}$

37. $3x^2 + 5x^2 = (3+5)x^2 = 8x^2$

39. $8x^5 - 19x^5 = (8-19)x^5 = -11x^5$

41. $2a + a - 3a = (2+1-3)a = 0a = 0$

43. $10x^3 - 8x^3 + 2x^3 = (10-8+2)x^3 = 4x^3$

45. $20ab^2 - 19ab^2 + 30ab^2 = (20-19+30)ab^2 = 31ab^2$

47. See the table in the back of the textbook.

49. $\dfrac{(3x^2)(8x^5)}{6x^4} = \dfrac{24x^7}{6x^4} = \dfrac{24}{6} \cdot \dfrac{x^7}{x^4} = 4x^3$

51. $\dfrac{(9a^2b)(2a^3b^4)}{18a^5b^7} = \dfrac{18a^5b^5}{18a^5b^7} = \dfrac{18}{18} \cdot \dfrac{a^5}{a^5} \cdot \dfrac{b^5}{b^7} = 1 \cdot \dfrac{1}{b^2} = \dfrac{1}{b^2}$

53. $\dfrac{(4x^3y^2)(9x^4y^{10})}{(3x^5y)(2x^6y)} = \dfrac{36x^7y^{12}}{6x^{11}y^2} = \dfrac{36}{6} \cdot \dfrac{x^7}{x^{11}} \cdot \dfrac{y^{12}}{y^2} = 6 \cdot \dfrac{1}{x^4} \cdot y^{10} = \dfrac{6y^{10}}{x^4}$

55. $\dfrac{(6 \times 10^8)(3 \times 10^5)}{9 \times 10^7} = \dfrac{18 \times 10^{13}}{9 \times 10^7} = 2 \times 10^6$

57. $\dfrac{(5 \times 10^3)(4 \times 10^{-5})}{2 \times 10^{-2}} = \dfrac{20 \times 10^{-2}}{2 \times 10^{-2}} = 10 = 1 \times 10^1$

© 2000 Harcourt, Inc

Problem Set 5.3

59. $\dfrac{(2.8 \times 10^{-7})(3.6 \times 10^{4})}{2.4 \times 10^{3}} = \dfrac{10.08 \times 10^{-3}}{2.4 \times 10^{3}} = 4.2 \times 10^{-6}$

61. $\dfrac{18x^{4}}{3x} + \dfrac{21x^{7}}{7x^{4}} = 6x^{3} + 3x^{3} = 9x^{3}$

63. $\dfrac{45a^{6}}{9a^{4}} - \dfrac{50a^{8}}{2a^{6}} = 5a^{2} - 25a^{2} = -20a^{2}$

65. $\dfrac{6x^{7}y^{4}}{3x^{2}y^{2}} + \dfrac{8x^{5}y^{8}}{2y^{6}} = 2x^{5}y^{2} + 4x^{5}y^{2} = 6x^{5}y^{2}$

67. $4^{x} \cdot 4^{5} = 4^{7}$
    $4^{x+5} = 4^{7}$
    $x + 5 = 7$
    $x = 2$

69. $(7^{3})^{x} = 7^{12}$
    $7^{3x} = 7^{12}$
    $3x = 12$
    $x = 4$

71. $(a+b)^{2} = (4+5)^{2} = 9^{2} = 81$
    $a^{2} + b^{2} = 4^{2} + 5^{2} = 16 + 25 = 41$
    Note that the values are not equal.

73. $(a+b)^{2} = (3+4)^{2} = 7^{2} = 49$
    $a^{2} + 2ab + b^{2} = 3^{2} + 2(3)(4) + 4^{2} = 9 + 24 + 16 = 49$
    Note that the values are equal.

75. Let x = width
    Let $2x$ = length
    $P = 2(x) + 2(2x) = 2x + 4x = 6x$
    $A = x \cdot 2x = 2x^{2}$

77. Let x = width
    Let $2x$ = length
    $V = (x)(2x)(4) = 8x^{2}$ inches^{3}

79. (a) $V = (8.5)(55)(10.1) = 4,700$ feet^{3} = 4.7×10^{3} feet^{3}
    (b) $n = \dfrac{4700}{0.15} = 31,000$ boxes = 3.1×10^{4} boxes

81. $y = x^{2}$ $(-4, 16), (-2, 4), (-1, 1), (0, 0), (1, 1), (2, 4), (4, 16)$
    See the graph in the back of the textbook.

83. $y = 2x^{2}$ $(-3, 18), (-2, 8), (-1, 2), (0, 0), (1, 2), (2, 8), (3, 18)$
    See the graph in the back of the textbook.

85. $y = \dfrac{1}{2}x^{2}$ $(-4, 8), (-2, 2), (-1, 1/2), (0, 0), (1, 1/2), (2, 2), (4, 8)$
    See the graph in the back of the textbook.

87. $y = \dfrac{1}{4}x^{2}$ $(-4, 4), (-2, 1), (-1, 1/4), (0, 0), (1, 1/4), (2, 1), (4, 4)$
    See the graph in the back of the textbook.

© 2000 Harcourt, Inc

89. Evaluating when $x = -2$: $-2x + 5 = -2(-2) + 5 = 4 + 5 = 9$

91. Evaluating when $x = -2$: $x^2 + 5x + 6 = (-2)^2 + 5(-2) + 6 = 4 - 10 + 6 = 0$

93. The ordered pairs are $(-2, -2)$, $(0, 2)$, and $(2, 6)$. See the graph in the back of the textbook.

95. The ordered pairs are $(-3, 0)$, $(0, 1)$, and $(3, 2)$. See the graph in the back of the textbook.

5.4 Addition and Subtraction of Polynomials

1. This is a trinomial of degree 3.
3. This is a trinomial of degree 3.
5. This is a trinomial of degree 1.
7. This is a binomial of degree 2.
9. This is a monomial of degree 2.
11. This is a monomial of degree 0.

13. $(2x^2 + 3x + 4) + (3x^2 + 2x + 5) = (2x^2 + 3x^2) + (3x + 2x) + (4 + 5) = 5x^2 + 5x + 9$

15. $(3a^2 - 4a + 1) + (2a^2 - 5a + 6) = (3a^2 + 2a^2) + (-4a - 5a) + (1 + 6) = 5a^2 - 9a + 7$

17. $x^2 + 4x + 2x + 8 = x^2 + (4x + 2x) + 8 = x^2 + 6x + 8$

19. $6x^2 - 3x - 10x + 5 = 6x^2 + (-3x - 10x) + 5 = 6x^2 - 13x + 5$

21. $x^2 - 3x + 3x - 9 = x^2 + (-3x + 3x) - 9 = x^2 - 9$

23. $3y^2 - 5y - 6y + 10 = 3y^2 + (-5y - 6y) + 10 = 3y^2 - 11y + 10$

25. $(6x^3 - 4x^2 + 2x) + (9x^2 - 6x + 3) = 6x^3 + (-4x^2 + 9x^2) + (2x - 6x) + 3 = 6x^3 + 5x^2 - 4x + 3$

27. $\left(\frac{2}{3}x^2 - \frac{1}{5}x - \frac{3}{4}\right) + \left(\frac{4}{3}x^2 - \frac{4}{5}x + \frac{7}{4}\right) = \left(\frac{2}{3}x^2 + \frac{4}{3}x^2\right) + \left(-\frac{1}{5}x - \frac{4}{5}x\right) + \left(-\frac{3}{4} + \frac{7}{4}\right) = 2x^2 - x + 1$

29. $(a^2 - a - 1) - (-a^2 + a + 1) = a^2 - a - 1 + a^2 - a - 1 = 2a^2 - 2a - 2$

31. $\left(\frac{5}{9}x^3 + \frac{1}{3}x^2 - 2x + 1\right) - \left(\frac{2}{3}x^3 + x^2 + \frac{1}{2}x - \frac{3}{4}\right) = \frac{5}{9}x^3 + \frac{1}{3}x^2 - 2x + 1 - \frac{2}{3}x^3 - x^2 - \frac{1}{2}x + \frac{3}{4}$

$= \left(\frac{5}{9} - \frac{2}{3}\right)x^3 + \left(\frac{1}{3} - 1\right)x^2 - \left(2 + \frac{1}{2}\right)x + 1 + \frac{3}{4}$

$= -\frac{1}{9}x^3 - \frac{2}{3}x^2 - \frac{5}{2}x + \frac{7}{4}$

33. $(4y^2 - 3y + 2) + (5y^2 + 12y - 4) - (13y^2 - 6y + 20)$

$= 4y^2 - 3y + 2 + 5y^2 + 12y - 4 - 13y^2 + 6y - 20$

$= (4y^2 + 5y^2 - 13y^2) + (-3y + 12y + 6y) + (2 - 4 - 20)$

$= -4y^2 + 15y - 22$

Problem Set 5.4

35. $(11x^2 - 10x + 13) - (10x^2 + 23x - 50) = 11x^2 - 10x + 13 - 10x^2 - 23x + 50$
$$= (11x^2 - 10x^2) + (-10x - 23x) + (13 + 50)$$
$$= x^2 - 33x + 63$$

37. $(11y^2 + 11y + 11) - (3y^2 + 7y - 15) = 11y^2 + 11y + 11 - 3y^2 - 7y + 15$
$$= (11y^2 - 3y^2) + (11y - 7y) + (11 + 15)$$
$$= 8y^2 + 4y + 26$$

39. $(25x^2 - 50x + 75) + (50x^2 - 100x - 150) = (25x^2 + 50x^2) + (-50x - 100x) + (75 - 150) = 75x^2 - 150x - 75$

41. $(3x - 2) + (11x + 5) - (2x + 1) = 3x - 2 + 11x + 5 - 2x - 1 = (3x + 11x - 2x) + (-2 + 5 - 1) = 12x + 2$

43. When $x = 3$: $x^2 - 2x + 1 = (3)^2 - 2(3) + 1 = 9 - 6 + 1 = 4$

45. When $y = 10$: $(y - 5)^2 = (10 - 5)^2 = (5)^2 = 25$

47. When $a = 2$: $a^2 + 4a + 4 = (2)^2 + 4(2) + 4 = 4 + 8 + 4 = 16$

49. $V_s = \frac{4}{3}\pi r^3 = \frac{4}{3}\pi (3)^3 = 36\pi$

$V_c = \pi r^2 h = \pi (3)^2 (6) = 54\pi$

Volume filled with padding $= V_c - V_s$

$V_c - V_s = 54\pi - 36\pi = 18\pi$

51. $3x(-5x) = -15x^2$

53. $2x(3x^2) = 6x^{1+2} = 6x^3$

55. $3x^2(2x^2) = 6x^{2+2} = 6x^4$

5.5 Multiplication with polynomials

1. $2x(3x+1) = 2x(3x) + 2x(1) = 6x^2 + 2x$
3. $2x^2(3x^2 - 2x + 1) = 2x^2(3x^2) + 2x^2(-2x) + 2x^2(1) = 6x^4 - 4x^3 + 2x^2$
5. $2ab(a^2 - ab + 1) = 2ab(a^2) - 2ab(ab) + 2ab(1) = 2a^3b - 2a^2b^2 + 2ab$
7. $y^2(3y^2 + 9y + 12) = 3y^4 + 9y^3 + 12y^2$
9. $4x^2y(2x^3y + 3x^2y^2 + 8y^3) = 4x^2y(2x^3y) + 4x^2y(3x^2y^2) + 4x^2y(8y^3) = 8x^5y^2 + 12x^4y^3 + 32x^2y^4$
11. Foil Method: $(x+3)(x+4) = x^2 + 3x + 4x + 12 = x^2 + 7x + 12$
13. Foil Method: $(x+6)(x+1) = x^2 + 6x + 1x + 6 = x^2 + 7x + 6$
15. Foil Method: $\left(x + \frac{1}{2}\right)\left(x + \frac{3}{2}\right) = x^2 + \frac{1}{2}x + \frac{3}{2}x + \frac{3}{4} = x^2 + 2x + \frac{3}{4}$
17. Foil Method: $(a+5)(a-3) = a^2 - 3a + 5a - 15 = a^2 + 2a - 15$
19. Foil Method: $(x-a)(y+b) = xy - ay + bx - ab$
21. Foil Method: $(x+6)(x-6) = x^2 + 6x - 6x - 36 = x^2 - 36$
23. Foil Method: $\left(y + \frac{5}{6}\right)\left(y - \frac{5}{6}\right) = y^2 + \frac{5}{6}y - \frac{5}{6}y - \frac{25}{36} = y^2 - \frac{25}{36}$
25. Foil Method: $(2x-3)(x-4) = 2x^2 - 3x - 8x + 12 = 2x^2 - 11x + 12$
27. Foil Method: $(a+2)(2a-1) = 2a^2 + 4a - a - 2 = 2a^2 + 3a - 2$
29. Foil Method: $(2x-5)(3x-2) = 6x^2 - 15x - 4x + 10 = 6x^2 - 19x + 10$
31. Foil Method: $(2x+3)(a+4) = 2ax + 3a + 8x + 12$
33. Foil Method: $(5x-4)(5x+4) = 25x^2 - 20x + 20x - 16 = 25x^2 - 16$
35. Foil Method: $\left(2x - \frac{1}{2}\right)\left(x + \frac{3}{2}\right) = 2x^2 - \frac{1}{2}x + 3x - \frac{3}{4} = 2x^2 + \frac{5}{2}x - \frac{3}{4}$
37. Foil Method: $(1-2a)(3-4a) = 3 - 6a - 4a + 8a^2 = 3 - 10a + 8a^2$
39. See the rectangle in the back of the textbook.
41. See the rectangle in the back of the textbook.

43. $a^2 - 3a + 2$
    $\underline{a - 3}$
    $a^3 - 3a^2 + 2a$
    $\underline{-3a^2 + 9a - 6}$
    $a^3 - 6a^2 + 11a - 6$

45. $x^2 - 2x + 4$
    $\underline{x + 2}$
    $x^3 - 2x^2 + 4x$
    $\underline{2x^2 - 4x + 8}$
    $x^3 + 8$

47. $x^2 + 8x + 9$
    $\underline{2x + 1}$
    $2x^3 + 16x^2 + 18x$
    $\underline{x^2 + 8x + 9}$
    $2x^3 + 17x^2 + 26x + 9$

49. $5x^2 + 2x + 1$
    $\underline{x^2 - 3x + 5}$
    $5x^4 + 2x^3 + x^2$
    $\underline{-15x^3 - 6x^2 - 3x}$
    $\underline{25x^2 + 10x + 5}$
    $5x^4 - 13x^3 + 20x^2 + 7x + 5$

Problem Set 5.5

51. $(x^2+3)(2x^2-5) = 2x^4 - 5x^2 + 6x^2 - 15 = 2x^4 + x^2 - 15$

53. $(3a^4+2)(2a^2+5) = 6a^6 + 15a^4 + 4a^2 + 10$

55. $(x+3)(x+4) = x^2 + 3x + 4x + 12 = x^2 + 7x + 12$

$$\begin{array}{r} x^2+7x+12 \\ \underline{x+5} \\ x^3+7x^2+12x \\ \underline{5x^2+35x+60} \\ x^3+12x^2+47x+60 \end{array}$$

57. Let x = width
Let $2x+5$ = length
$A = x(2x+5) = 2x^2 + 5x$

59. Let x and $x+1$ = width and length, respectively.
$A = x(x+1) = x^2 + x$

61. $x = 1200 - 100p$
$R = xp = (1200 - 100p)p$
$R = 1200p - 100p^2$

63. $x = 1700 - 100p$
$R = xp = (1700 - 100p)p$
$R = 1700p - 100p^2$

65. The intersection point is $(3,1)$.
See the graph in the back of the textbook.

67. Multiply the first equation by 3 and the second equation by -2:
$6x + 9y = -3$
$-6x - 10y = 4$
Adding the two equations:
$-y = 1$
$y = -1$
Substituting into the first equation:
$2x + 3(-1) = -1$
$2x - 3 = -1$
$2x = 2$
$x = 1$
The solution is $(1, -1)$.

69. Substituting $y = 3x + 1$ from the second equation into the first equation:
$2x - 6(3x+1) = 2$
$2x - 18x - 6 = 2$
$-16x - 6 = 2$
$-16x = 8$
$x = -\dfrac{1}{2}$
Substituting into the second equation:
$y = 3\left(-\dfrac{1}{2}\right) + 1 = -\dfrac{3}{2} + 1 = -\dfrac{1}{2}$
The solution is $\left(-\dfrac{1}{2}, -\dfrac{1}{2}\right)$.

71. Let x = the number of dimes
Let y = the number of quarters
$x + y = 11$
$0.10x + 0.25y = 1.85$
Multiplying the first equation by -0.10:

71. $-0.10x - 0.10y = -1.10$
$0.10x + 0.25y = 1.85$
Adding the two equations:
$0.15y = 0.75$
$y = 5$
Substituting into the first equation:
$x + 5 = 11$
$x = 6$
Amy has 6 dimes and 5 quarters.

5.6 Binomial Squares and Other Special Products

1. Foil Method: $(x-2)^2 = (x-2)(x-2) = x^2 - 2x - 2x + 4 = x^2 - 4x + 4$

3. Foil Method: $(a+3)^2 = (a+3)(a+3) = a^2 + 3a + 3a + 9 = a^2 + 6a + 9$

5. Foil Method: $(x-5)^2 = (x-5)(x-5) = x^2 - 5x - 5x + 25 = x^2 - 10x + 25$

7. Foil Method: $\left(a - \frac{1}{2}\right)^2 = \left(a - \frac{1}{2}\right)\left(a - \frac{1}{2}\right) = a^2 - \frac{1}{2}a - \frac{1}{2}a + \frac{1}{4} = a^2 - a + \frac{1}{4}$

9. Foil Method: $(x+10)^2 = (x+10)(x+10) = x^2 + 10x + 10x + 100 = x^2 + 20x + 100$

11. $(a+0.8)^2 = a^2 + 2(a)(0.8) + (0.8)^2 = a^2 + 1.6a + 0.64$

13. $(2x-1)^2 = (2x)^2 - 2(2x)(1) + (1)^2 = 4x^2 - 4x + 1$

15. $(4a+5)^2 = (4a)^2 + 2(4a)(5) + (5)^2 = 16a^2 + 40a + 25$

17. $(3x-2)^2 = 9x^2 + 2(3x)(-2) + 4 = 9x^2 - 12x + 4$

19. $(3a+5b)^2 = (3a)^2 + 2(3a)(5b) + (5b)^2 = 9a^2 + 30ab + 25b^2$

21. $(4x-5y)^2 = 16x^2 + 2(4x)(-5y) + 25y^2 = 16x^2 - 40xy + 25y^2$

23. $(7m+2n)^2 = (7m)^2 + 2(7m)(2n) + (2n)^2 = 49m^2 + 28mn + 4n^2$

25. $(6x-10y)^2 = 36x^2 + 2(6x)(-10y) + 100y^2 = 36x^2 - 120xy + 100y^2$

27. $(x^2+5)^2 = (x^2)^2 + 2(x^2)(5) + (5)^2 = x^4 + 10x^2 + 25$

29. $(a^2+1)^2 = a^4 + 2(a^2)(1) + 1^2 = a^4 + 2a^2 + 1$

31. See the table in the back of the textbook.

33. See the table in the back of the textbook.

35. $(a+5)(a-5) = a^2 + 5a - 5a - 25 = a^2 - 25$

37. $(y-1)(y+1) = y^2 - y + y - 1 = y^2 - 1$

39. $(9+x)(9-x) = (9)^2 - (x)^2 = 81 - x^2$

41. $(2x+5)(2x-5) = (2x)^2 - (5)^2 = 4x^2 - 25$

43. $\left(4x + \frac{1}{3}\right)\left(4x - \frac{1}{3}\right) = (4x)^2 - \left(\frac{1}{3}\right)^2 = 16x^2 - \frac{1}{9}$

45. $(2a+7)(2a-7) = (2a)^2 - (7)^2 = 4a^2 - 49$

47. $(6-7x)(6+7x) = (6)^2 - (7x)^2 = 36 - 49x^2$

49. $(x^2+3)(x^2-3) = (x^2)^2 - (3)^2 = x^4 - 9$

51. $(a^2+4)(a^2-4) = (a^2)^2 - (4)^2 = a^4 - 16$

53. $(5y^4-8)(5y^4+8) = (5y^4)^2 - (8)^2 = 25y^8 - 64$

Problem Set 5.6

55. $(x+3)(x-3)+(x+5)(x-5)=(x^2-9)+(x^2-25)=2x^2-34$

57. $(2x+3)^2-(4x-1)^2=(4x^2+12x+9)-(16x^2-8x+1)=4x^2+12x+9-16x^2+8x-1=-12x^2+20x+8$

59. $(a+1)^2-(a+2)^2+(a+3)^2=(a^2+2a+1)-(a^2+4a+4)+(a^2+6a+9)$
$$=a^2+2a+1-a^2-4a-4+a^2+6a+9$$
$$=a^2+4a+6$$

61. $(2x+3)^3=(2x+3)(2x+3)^2$
$$=(2x+3)(4x^2+12x+9)$$
$$=8x^3+24x^2+18x+12x^2+36x+27$$
$$=8x^3+36x^2+54x+27$$

63. $49(51)=(50-1)(50+1)=(50)^2-(1)^2=2,500-1=2,499$

65. $(x+3)^2=(2+3)^2=(5)^2=25$
$x^2+6x+9=(2)^2+6(2)+9=4+12+9=25$

67. Let x and $x+1=$ the two consecutive integers.
$(x)^2+(x+1)^2=x^2+(x^2+2x+1)=2x^2+2x+1$

69. Let x, $x+1$ and $x+2=$ the three consecutive integers.
$(x)^2+(x+1)^2+(x+2)^2=x^2+(x^2+2x+1)+(x^2+4x+4)=3x^2+6x+5$

71. Verifying the areas: $(a+b)^2=a^2+ab+ab+b^2=a^2+2ab+b^2$

73. See the rectangle in the back of the textbook.

75. $\dfrac{15x^2y}{3xy}=\dfrac{15}{3}\cdot\dfrac{x^2}{x}\cdot\dfrac{y}{y}=5x$

77. $\dfrac{35a^6b^8}{70a^2b^{10}}=\dfrac{35}{70}\cdot\dfrac{a^6}{a^2}\cdot\dfrac{b^8}{b^{10}}=\dfrac{1}{2}\cdot a^4\cdot\dfrac{1}{b^2}=\dfrac{a^4}{2b^2}$

79. The intersection point is $(3,-1)$.
See the graph in the back of the textbook.

81. The intersection point is $(-1, 1)$.
See the graph in the back of the textbook.

© 2000 Harcourt, Inc

5.7 Dividing a Polynomial by a Monomial

1. $\dfrac{5x^2 - 10x}{5x} = \dfrac{5x^2}{5x} - \dfrac{10x}{5x} = x - 2$

3. $\dfrac{15x - 10x^3}{5x} = \dfrac{15x}{5x} - \dfrac{10x^3}{5x} = 3 - 2x^2$

5. $\dfrac{25x^2 y - 10xy}{5x} = \dfrac{25x^2 y}{5x} - \dfrac{10xy}{5x} = 5xy - 2y$

7. $\dfrac{35x^5 - 30x^4 + 25x^3}{5x} = \dfrac{35x^5}{5x} - \dfrac{30x^4}{5x} + \dfrac{25x^3}{5x} = 7x^4 - 6x^3 + 5x^2$

9. $\dfrac{50x^5 - 25x^3 + 5x}{5x} = \dfrac{50x^5}{5x} - \dfrac{25x^3}{5x} + \dfrac{5x}{5x} = 10x^4 - 5x^2 + 1$

11. $\dfrac{8a^2 - 4a}{-2a} = \dfrac{8a^2}{-2a} + \dfrac{-4a}{-2a} = -4a + 2$

13. $\dfrac{16a^5 + 24a^4}{-2a} = \dfrac{16a^5}{-2a} + \dfrac{24a^4}{-2a} = -8a^4 - 12a^3$

15. $\dfrac{8ab + 10a^2}{-2a} = \dfrac{8ab}{-2a} + \dfrac{10a^2}{-2a} = -4b - 5a$

17. $\dfrac{12a^3 b - 6a^2 b^2 + 14ab^3}{-2a} = \dfrac{12a^3 b}{-2a} - \dfrac{6a^2 b^2}{-2a} + \dfrac{14ab^3}{-2a} = -6a^2 b + 3ab^2 - 7b^3$

19. $\dfrac{a^2 + 2ab + b^2}{-2a} = \dfrac{a^2}{-2a} + \dfrac{2ab}{-2a} + \dfrac{b^2}{-2a} = -\dfrac{a}{2} - b - \dfrac{b^2}{2a}$

21. $\dfrac{6x + 8y}{2} = \dfrac{6x}{2} + \dfrac{8y}{2} = 3x + 4y$

23. $\dfrac{7y - 21}{-7} = \dfrac{7y}{-7} + \dfrac{-21}{-7} = -y + 3$

25. $\dfrac{10xy - 8x}{2x} = \dfrac{10xy}{2x} - \dfrac{8x}{2x} = 5y - 4$

27. $\dfrac{x^2 y - x^3 y^2}{x} = \dfrac{x^2 y}{x} - \dfrac{x^3 y^2}{x} = xy - x^2 y^2$

29. $\dfrac{x^2 y - x^3 y^2}{-x^2 y} = \dfrac{x^2 y}{-x^2 y} - \dfrac{x^3 y^2}{-x^2 y} = -1 + xy$

31. $\dfrac{a^2 b^2 - ab^2}{-ab^2} = \dfrac{a^2 b^2}{-ab^2} + \dfrac{-ab^2}{-ab^2} = -a + 1$

33. $\dfrac{x^3 - 3x^2 y + xy^2}{x} = \dfrac{x^3}{x} - \dfrac{3x^2 y}{x} + \dfrac{xy^2}{x} = x^2 - 3xy + y^2$

35. $\dfrac{10a^2 - 15a^2 b + 25a^2 b^2}{5a^2} = \dfrac{10a^2}{5a^2} - \dfrac{15a^2 b}{5a^2} + \dfrac{25a^2 b}{5a^2} = 2 - 3b + 5b$

Problem Set 5.7

37. $\dfrac{26x^2y^2 - 13xy}{-13xy} = \dfrac{26x^2y^2}{-13xy} - \dfrac{13xy}{-13xy} = -2xy + 1$

39. $\dfrac{4x^2y^2 - 2xy}{4xy} = \dfrac{4x^2y^2}{4xy} - \dfrac{2xy}{4xy} = xy - \dfrac{1}{2}$

41. $\dfrac{5a^2x - 10ax^2 + 15a^2x^2}{20a^2x^2} = \dfrac{5a^2x}{20a^2x^2} - \dfrac{10ax^2}{20a^2x^2} + \dfrac{15a^2x^2}{20a^2x^2} = \dfrac{1}{4x} - \dfrac{1}{2a} + \dfrac{3}{4}$

43. $\dfrac{16x^5 + 8x^2 + 12x}{12x^3} = \dfrac{16x^5}{12x^3} + \dfrac{8x^2}{12x^3} + \dfrac{12x}{12x^3} = \dfrac{4x^2}{3} + \dfrac{2}{3x} + \dfrac{1}{x^2}$

45. $\dfrac{9a^{5m} - 27a^{3m}}{3a^{2m}} = \dfrac{9a^{5m}}{3a^{2m}} - \dfrac{27a^{3m}}{3a^{2m}} = 3a^{5m-2m} - 9a^{3m-2m} = 3a^{3m} - 9a^{m}$

47. $\dfrac{10x^{5m} - 25x^{3m} + 35x^{m}}{5x^{m}} = \dfrac{10x^{5m}}{5x^{m}} - \dfrac{25x^{3m}}{5x^{m}} + \dfrac{35x^{m}}{5x^{m}}\ 2x^{5m-m} - 5x^{3m-m} + 7x^{m-m} = 2x^{4m} - 5x^{2m} + 7$

49. $\dfrac{2x^3(3x+2) - 3x^2(2x-4)}{2x^2} = \dfrac{6x^4 + 4x^3 - 6x^3 + 12x^2}{2x^2}$

$= \dfrac{6x^4}{2x^2} + \dfrac{4x^3}{2x^2} - \dfrac{6x^3}{2x^2} + \dfrac{12x^2}{2x^2}$

$= 3x^2 + 2x - 3x + 6$

$= 3x^2 - x + 6$

51. $\dfrac{(x+2)^2 - (x-2)^2}{2x} = \dfrac{(x^2+4x+4) - (x^2-4x+4)}{2x} = \dfrac{x^2+4x+4-x^2+4x-4}{2x} = \dfrac{8x}{2x} = 4$

53. $\dfrac{(x+5)^2 + (x+5)(x-5)}{2x} = \dfrac{x^2+10x+25+x^2-25}{2x} = \dfrac{2x^2+10x}{2x} = \dfrac{2x^2}{2x} + \dfrac{10x}{2x} = x+5$

55. When $x = 2$:

$\dfrac{10x+15}{5} = \dfrac{10(2)+15}{5} = \dfrac{20+15}{5} = \dfrac{35}{5} = 7$

$2x+3 = 2(2)+3 = 4+3 = 7$

57. When $x = 10$:

$\dfrac{3x+8}{2} = \dfrac{3(10)+8}{2} = \dfrac{30+8}{2} = \dfrac{38}{2} = 19$

$3x+4 = 3(10)+4 = 30+4 = 34$

Thus, $\dfrac{3x+8}{2} \neq 3x+4$.

© 2000 Harcourt, Inc

59. Adding the two equations:
    $$2x = 14$$
    $$x = 7$$
    Substituting into the first equation:
    $$7 + y = 6$$
    $$y = -1$$
    The solution is $(7, -1)$.

61. Multiplying the second equation by 3:
    $$2x - 3y = -5$$
    $$3x + 3y = 15$$
    Adding the two equations:
    $$5x = 10$$
    $$x = 2$$
    Substituting into the second equation:
    $$2 + y = 5$$
    $$y = 3$$
    The solution is $(2, 3)$.

63. Substituting y from the second equation into the first equation:
    $$x + 2x - 1 = 2$$
    $$3x - 1 = 2$$
    $$3x = 3$$
    $$x = 1$$
    Substituting into the second equation:
    $$y = 2(1) - 1 = 2 - 1 = 1$$
    The solution is $(1, 1)$.

65. Substituting y from the second equation into the first equation:
    $$4x + 2(-2x + 4) = 8$$
    $$4x - 4x + 8 = 8$$
    $$8 = 8$$
    Since this statement is true, the system is dependent. The two lines coincide.

5.8 Dividing a Polynomial by a Polynomial

1.
$$\begin{array}{r} x - 2 \\ x-3 \overline{) x^2 - 5x + 6} \\ \underline{x^2 - 3x} \\ -2x + 6 \\ \underline{-2x + 6} \\ 0 \end{array}$$
The answer is $x - 2$

3.
$$\begin{array}{r} a + 4 \\ a+5 \overline{) a^2 + 9a + 20} \\ \underline{a^2 + 5a} \\ 4a + 20 \\ \underline{4a + 20} \\ 0 \end{array}$$
The answer is $a + 4$

5.
$$\begin{array}{r} x - 3 \\ x-3 \overline{) x^2 - 6x + 9} \\ \underline{x^2 - 3x} \\ -3x + 9 \\ \underline{-3x + 9} \\ 0 \end{array}$$
The answer is $x - 3$

7.
$$\begin{array}{r} x + 3 \\ 2x-1 \overline{) 2x^2 + 5x - 3} \\ \underline{2x^2 - x} \\ 6x - 3 \\ \underline{6x - 3} \\ 0 \end{array}$$
The answer is $x + 3$

Problem Set 5.8

9.
$$\begin{array}{r} a-5 \\ 2a+1 \overline{\smash{\big)}\ 2a^2-9a-5} \\ \underline{2a^2+a} \\ -10a-5 \\ \underline{-10a-5} \\ 0 \end{array}$$

The answer is $a-5$

11.
$$\begin{array}{r} x+2 \\ x+3 \overline{\smash{\big)}\ x^2+5x+8} \\ \underline{x^2+3x} \\ 2x+8 \\ \underline{2x+6} \\ 2 \end{array}$$

The answer is $x+2+\dfrac{2}{x+3}$

13.
$$\begin{array}{r} a-2 \\ a+5 \overline{\smash{\big)}\ a^2+3a+2} \\ \underline{a^2+5a} \\ -2a+2 \\ \underline{-2a-10} \\ 12 \end{array}$$

The answer is $a-2+\dfrac{12}{a+5}$

15.
$$\begin{array}{r} x+4 \\ x-2 \overline{\smash{\big)}\ x^2+2x+1} \\ \underline{x^2-2x} \\ 4x+1 \\ \underline{4x-8} \\ 9 \end{array}$$

The answer is $x+4+\dfrac{9}{x-2}$

17.
$$\begin{array}{r} x+4 \\ x+1 \overline{\smash{\big)}\ x^2+5x-6} \\ \underline{x^2+x} \\ 4x-6 \\ \underline{4x+4} \\ -10 \end{array}$$

The answer is $x+4-\dfrac{10}{x+1}$

19.
$$\begin{array}{r} a+1 \\ a+2 \overline{\smash{\big)}\ a^2+3a+1} \\ \underline{a^2+2a} \\ a+1 \\ \underline{a+2} \\ -1 \end{array}$$

The answer is $a+1+\dfrac{-1}{a+2}$

21.
$$\begin{array}{r} x-3 \\ 2x+4 \overline{\smash{\big)}\ 2x^2-2x+5} \\ \underline{2x^2+4x} \\ -6x+5 \\ \underline{-6x-12} \\ 17 \end{array}$$

The answer is $x-3+\dfrac{17}{2x+4}$

23.
$$\begin{array}{r} 3a-2 \\ 2a+3 \overline{\smash{\big)}\ 6a^2+5a+1} \\ \underline{6a^2+9a} \\ -4a+1 \\ \underline{-4a-6} \\ 7 \end{array}$$

The answer is $3a-2+\dfrac{7}{2a+3}$

© 2000 Harcourt, Inc

25.
$$\begin{array}{r}2a^2-a-3\\3a-5{\overline{\smash{\big)}\,6a^3-13a^2-4a+15}}\\\underline{6a^3-10a^2}\\-3a^2-4a\\\underline{-3a^2+5a}\\-9a+15\\\underline{-9a+15}\\0\end{array}$$

The answer is $2a^2-a-3$

27.
$$\begin{array}{r}x^2-x+5\\x+1{\overline{\smash{\big)}\,x^3+0x^2+4x+5}}\\\underline{x^3+x^2}\\-x^2+4x+5\\\underline{-x^2-x}\\5x+5\\\underline{5x+5}\\0\end{array}$$

The answer is x^2-x+5

29.
$$\begin{array}{r}x^2+x+1\\x-1{\overline{\smash{\big)}\,x^3+0x^2+0x-1}}\\\underline{x^3-x^2}\\x^2+0x\\\underline{x^2-x}\\x-1\\\underline{x-1}\\0\end{array}$$

The answer is x^2+x+1

31.
$$\begin{array}{r}x^2+2x+4\\x-2{\overline{\smash{\big)}\,x^3+0x^2+0x-8}}\\\underline{x^3-2x^2}\\2x^2+0x\\\underline{2x^2-4x}\\4x-8\\\underline{4x-8}\\0\end{array}$$

The answer is x^2+2x+4

33. Let x and $y =$ the two numbers

$x+y=25$

$y=4x$

Substituting into the first equation:

$x+4x=25$

$5x=25$

$x=5$

$y=4(5)=20$

The two numbers are 5 and 20.

35. Let $x =$ amount invested at 8%

Let $y =$ amount invested at 9%

$x+y=1200$ (1)

$0.08x+0.09y=100$ (2)

Solve (1) for x:

$x=1200-y$ (3)

Substituting (3) into (2)

$0.08(1200-y)+0.09y=100$

$96-0.08y+0.09y=100$

$96+0.01y=100$

$0.01y=4$

$y=400$

Substitute $y=400$ into (3)

$x=1200-400=800$

You invested $800 at 8% and $400 at 9%.

37. Let x = number of $5 bills
    Let y = number of $10 bills
    $$y = x + 4 \quad (1)$$
    $$5x + 10y = 160 \quad (2)$$
    Substitute (1) into (2)
    $$5x + 10(x + 4) = 160$$
    $$5x + 10x + 40 = 160$$
    $$15x + 40 = 160$$
    $$15x = 120$$
    $$x = 8$$
    Substitute $x = 8$ into (1)
    $$y = 8 + 4 = 12$$
    You have 8 $5 bills and 12 $10 bills.

39. Let x = gallons of 20% solution
    Let y = gallons of 60% solution
    $$x + y = 16 \quad (1)$$
    $$0.20x + 0.60y = 16(0.35) \quad (2)$$
    Solve (1) for x:
    $$x = 16 - y \quad (3)$$
    Substituting (3) into (2)
    $$0.20(16 - y) + 0.60y = 5.6$$
    $$3.2 - 0.20j + 0.60y = 5.6$$
    $$3.2 + 0.4y = 5.6$$
    $$0.4y = 2.4$$
    $$y = 6$$
    Substitute $y = 6$ into (3)
    $$x = 16 - 6 = 10$$
    10 gallons of 20% antifreeze solution must be mixed with 6 gallons of 60% antifreeze solution.

Chapter 5 Review

1. $(-1)^3 = (-1)(-1)(-1) = -1$

3. $\left(\frac{3}{7}\right)^2 = \left(\frac{3}{7}\right)\left(\frac{3}{7}\right) = \frac{9}{49}$

5. $x^{15} \cdot x^7 \cdot x^5 \cdot x^3 = x^{15+7+5+3} = x^{30}$

7. $(2^6)^4 = 2^{6 \cdot 4} = 2^{24}$

9. $(-2xyz)^3 = (-2)^3 x^3 y^3 z^3 = -8x^3 y^3 z^3$

11. $4x^{-5} = 4\left(\frac{1}{x^5}\right) = \frac{4}{x^5}$

13. $\dfrac{a^9}{a^3} = a^{9-3} = a^6$

15. $\dfrac{x^9}{x^{-6}} = x^{9-(-6)} = x^{15}$

17. $(-3xy)^0 = 1$

19. $\left(3x^3 y^2\right)^2 = 3^2 \left(x^3\right)^2 \left(y^2\right)^2 = 9x^{3(2)} y^{2(2)} = 9x^6 y^4$

21. $(-3xy^2)^{-3} = \dfrac{1}{(-3xy^2)^3}$
    $$= \dfrac{1}{(-3)^3 x^3 (y^2)^3}$$
    $$= \dfrac{1}{-27 x^3 y^{2(3)}}$$
    $$= \dfrac{-1}{27 x^3 y^6}$$

23. $\dfrac{(x^{-3})^3(x^6)^{-1}}{(x^{-5})^{-4}} = \dfrac{x^{-3(3)}x^{6(-1)}}{x^{-5(-4)}}$

$= \dfrac{x^{-9}x^{-6}}{x^{20}}$

$= \dfrac{x^{-15}}{x^{20}}$

$= \dfrac{1}{x^{20}x^{15}}$

$= \dfrac{1}{x^{35}}$

25. $\dfrac{(10x^3y^5)(21x^2y^6)}{(7xy^3)(5x^9y)} = \dfrac{10 \cdot 21x^3x^2y^5y^6}{7 \cdot 5xx^9y^3y}$

$= \dfrac{210x^{3+2}y^{5+6}}{35x^{1+9}y^{3+1}}$

$= \dfrac{6x^5y^{11}}{x^{10}y^4}$

$= 6x^{5-10}y^{11-4}$

$= 6x^{-5}y^7$

$= \dfrac{6y^7}{x^5}$

27. $\dfrac{8x^8y^3}{2x^3y} - \dfrac{10x^6y^9}{5xy^7} = 4x^{8-3}y^{3-1} - 2x^{6-1}y^{9-7}$

$= 4x^5y^2 - 2x^5y^2$

$= (4-2)x^5y^2$

$= 2x^5y^2$

29. $\dfrac{4.6 \times 10^5}{2 \times 10^{-3}} = 2.3 \times 10^{5-(-3)}$

$= 2.3 \times 10^8$

31. $(3a^2 - 5a + 5) + (5a^2 - 7a - 8) = 3a^2 + 5a^2 - 5a - 7a + 5 - 8$

$= 8a^2 - 12a - 3$

33. $(4x^2 - 3x - 2) - (8x^2 + 3x - 2) = 4x^2 - 3x - 2 - 8x^2 - 3x + 2$

$= -4x^2 - 6x$

35. $3x(4x - 7) = 3x(4x) + 3x(-7) = 12x^2 - 21x$

37. $\begin{array}{r} a^2 + 5a - 4 \\ a + 1 \\ \hline a^3 + 5a^2 - 4a \\ a^2 + 5a - 4 \\ \hline a^3 + 6a^2 + a - 4 \end{array}$

39. $(3x - 7)(2x - 5) = 3x(2x) + 3x(-5) + (-7)(2x) + (-7)(-5)$

$= 6x^2 - 15x - 14x + 35$

$= 6x^2 - 29x + 35$

41. $(a^2 - 3)(a^2 + 3) = (a^2)^2 - (3)^2 = a^4 - 9$

43. $(3x + 4)^2 = (3x)^2 + 2(3x)(4) + 4^2 = 9x^2 + 24x + 16$

45. $\dfrac{10ab}{-5a} + \dfrac{20a^2}{-5a} = -2b - 4a$

Chapter 5 Review

47.
$$\begin{array}{r} x+9 \\ x+6 \overline{)x^2+15x+54} \\ \underline{x^2+6x} \\ 9x+54 \\ \underline{9x+54} \\ 0 \end{array}$$

49.
$$\begin{array}{r} x^2-4x+16 \\ x+4 \overline{)x^3+0x^2+0x+64} \\ \underline{x^3+4x^2} \\ -4x^2+0x \\ \underline{-4x^2-16x} \\ 16x+64 \\ \underline{16x+64} \\ 0 \end{array}$$

51.
$$\begin{array}{r} x^2-4x+5 \\ 2x+1 \overline{)2x^3-7x^2+6x+10} \\ \underline{2x^3+x^2} \\ -8x^2+6x \\ \underline{-8x^2-4x} \\ 10x+10 \\ \underline{10x+5} \\ 5 \end{array}$$

53. $V = 3x^3$, $x = 2$

$V = 3(2)^3 = 3(8) = 24$ cubic feet.

Yes, the volume is greater than the volume of your refrigerator.

55. $A = s^2$

$225 = s^2$

$15 = s$

Maximum diameter = 15 feet

Maximum radius = 7 1/2 feet

The longest trampoline is the 6 foot radius

© 2000 Harcourt, Inc

Cumulative Review: Chapters 1-5

1. $-\left(-\dfrac{3}{4}\right) = \dfrac{3}{4}$

3. $6 \cdot 7 + 7 \cdot 9 = 42 + 63 = 105$

5. $6(4a+2) - 3(5a-1) = 24a + 12 - 15a + 3 = 24a - 15a + 12 + 3 = 9a + 15$

7. $-15 - (-3) = -15 + 3 = -12$

9. $(-9)(-5) = 45$

11. $(2y)^4 = 2^4 y^4 = 16y^4$

13. $\dfrac{(12xy^5)^3 (16x^2 y^2)}{(8x^3 y^3)(3x^5 y)} = \dfrac{1728 x^3 y^{15} \cdot 16 x^2 y^2}{24 x^8 y^4} = \dfrac{27648 x^5 y^{17}}{24 x^8 y^4} = \dfrac{1152 y^{13}}{x^3}$

15. $(5x-1)^2 = (5x)^2 - 2(5x)(1) + (1)^2 = 25x^2 - 10x + 1$

17. $\begin{array}{r} x^2 + x + 1 \\ \underline{x - 1} \\ x^3 + x^2 + x \\ \underline{-x^2 - x - 1} \\ x^3 - 1 \end{array}$

19. $6a - 5 = 4a$
    $-5 = -2a$
    $a = \dfrac{5}{2}$

21. $2(3x+5) + 8 = 2x + 10$
    $6x + 10 + 8 = 2x + 10$
    $6x + 18 = 2x + 10$
    $4x + 18 = 10$
    $4x = -8$
    $x = -2$

23. $3(2t-5) - 7 \leq 5(3t+1) + 5$
    $6t - 15 - 7 \leq 15t + 5 + 5$
    $6t - 22 \leq 15t + 10$
    $-9t - 22 \leq 10$
    $-9t \leq 32$
    $t \geq -32/9$
    See the graph in the back of the textbook

25. $-5 \leq 4x + 3 \leq 11$
    $-8 \leq 4x \leq 8$
    $-2 \leq x \leq 2$
    See the graph in the back of the textbook.

27. $\dfrac{15x^5 - 10x^2 + 20x}{5x^5} = \dfrac{15x^5}{5x^5} - \dfrac{10x^2}{5x^5} + \dfrac{20x}{5x^5} = 3 - \dfrac{2}{x^3} + \dfrac{4}{x^4}$

29. See the graph in the back of the textbook.

31. The intersection point is $(0, 1)$.
    See the graph in the back of the textbook.

33. The system is dependent. The two lines coincide.
    See the graph in the back of the textbook.

© 2000 Harcourt, Inc

Cumulative Review: Chapters 1-5

35. Multiplying the first equation by -3:
$$-3x - 6y = -15$$
$$3x + 6y = 14$$
Adding the two equations:
$$0 = -1$$
Since this statement is false, there is no solution to the system. The two lines are parallel.

37. Multiply the first equation by 12 and the second equation by 5:
$$12\left(\frac{1}{6}x + \frac{1}{4}y\right) = 12(1) \qquad 5\left(\frac{6}{5}x - y\right) = 5\left(\frac{8}{5}\right)$$
$$2x + 3y = 12 \qquad 6x - 5y = 8$$
The system of equation is:
$$2x + 3y = 12$$
$$6x - 5y = 8$$
Multiplying the first equation by -3:
$$-6x - 9y = -36$$
$$6x - 5y = 8$$
Adding the two equations:
$$-14y = -28$$
$$y = 2$$
Substituting into $2x + 3y = 12$:
$$2x + 3(2) = 12$$
$$2x + 6 = 12$$
$$2x = 6$$
$$x = 3$$
The solution is $(3, 2)$.

39. Solving the second equation for x:
$$2y = x - 3$$
$$x = 2y + 3$$
Substituting into the first equation:
$$4(2y + 3) + 5y = 25$$
$$8y + 12 + 5y = 25$$
$$13y + 12 = 25$$
$$13y = 13$$
$$y = 1$$
$$x = 2(1) + 3 = 5$$
The solution is $(5, 1)$.

41. When $x = 3$: $8x - 3 = 8(3) - 3 = 24 - 3 = 21$

43. The irrational numbers are $-\sqrt{2}$ and π.

45. Substituting each ordered pair:
$(0, 3)$: $3(0) - 4(3) = 0 - 12 = -12 \neq 12$
$(4, 0)$: $3(4) - 4(0) = 12 - 0 = 12$
$\left(\frac{16}{3}, 1\right)$: $3\left(\frac{16}{3}\right) - 4(1) = 16 - 4 = 12$
The ordered pairs $(4, 0)$ and $\left(\frac{16}{3}, 1\right)$ are solutions to the equation.

© 2000 Harcourt, Inc

47. $m = \dfrac{-1-3}{6-(-2)} = \dfrac{-4}{8} = -\dfrac{1}{2}$

Using the point-slope formula:
$$y - 3 = -\dfrac{1}{2}(x-(-2))$$
$$y - 3 = -\dfrac{1}{2}(x+2)$$
$$y - 3 = -\dfrac{1}{2}x - 1$$
$$y = -\dfrac{1}{2}x + 2$$

47. To find the x-intercept, let $y = 0$:
$$0 = -\dfrac{1}{2}x + 2$$
$$-2 = -\dfrac{1}{2}x$$
$$x = 4$$

To find the y-intercept, let $x = 0$:
$$y = -\dfrac{1}{2}(0) + 2 = 0 + 2 = 2$$

x-intercept is 4.
y-intercept is 2.

49. $m = \dfrac{8-4}{1-(-2)} = \dfrac{4}{3}$

Using the point-slope formula:
$$y - 8 = \dfrac{4}{3}(x-1)$$
$$y - 8 = \dfrac{4}{3}x - \dfrac{4}{3}$$
$$y = \dfrac{4}{3}x - \dfrac{4}{3} + 8$$
$$y = \dfrac{4}{3}x - \dfrac{4}{3} + \dfrac{24}{3}$$

Chapter 5 Test

1. $(-3)^4 = (-3)(-3)(-3)(-3) = 81$

2. $\left(\dfrac{3}{4}\right)^2 = \left(\dfrac{3}{4}\right)\left(\dfrac{3}{4}\right) = \dfrac{9}{16}$

3. $(3x^3)^2(2x^4)^3 = 3^2 \cdot x^{3 \cdot 2} \cdot 2^3 \cdot x^{4 \cdot 3}$
   $= 9x^6 \cdot 8x^{12}$
   $= 9 \cdot 8 x^{6+12}$
   $= 72x^{18}$

4. $3^{-2} = \dfrac{1}{3^2} = \dfrac{1}{9}$

5. $(3a^4 b^2)^0 = 1$

6. $\dfrac{a^{-3}}{a^{-5}} = a^{-3-(-5)}$
   $= a^{-3+5}$
   $= a^2$

Chapter 5 Test

7. $\dfrac{(x^{-2})^3 (x^{-3})^{-5}}{(x^{-4})^{-2}} = \dfrac{x^{-2(3)} x^{-3(-5)}}{x^{-4(-2)}}$

$= \dfrac{x^{-6} x^{15}}{x^8}$

$= \dfrac{x^9}{x^8}$

$= x$

8. $0.0278 = 2.78 \times 10^{-2}$

9. $2.43 \times 10^5 = 2.43 \times 100{,}000 = 243{,}000$

10. $\dfrac{35 x^2 y^4 z}{70 x^6 y^2 z} = \dfrac{35}{70} \cdot \dfrac{x^2}{x^6} \cdot \dfrac{y^4}{y^2} \cdot \dfrac{z}{z}$

$= \dfrac{1}{2} \cdot \dfrac{1}{x^4} \cdot \dfrac{y^2}{1} \cdot 1$

$= \dfrac{y^2}{2x^4}$

11. $\dfrac{(6a^2 b)(9a^3 b^2)}{18 a^4 b^3} = \dfrac{54 a^5 b^3}{18 a^4 b^3}$

$= \dfrac{54}{18} \cdot \dfrac{a^5}{a^4} \cdot \dfrac{b^3}{b^3}$

$= \dfrac{3}{1} \cdot \dfrac{a}{1} \cdot 1$

$= 3a$

12. $\dfrac{24 x^7}{3 x^2} + \dfrac{14 x^9}{7 x^4} = 8x^5 + 2x^5 = 10 x^5$

13. $\dfrac{(2.4 \times 10^5)(4.5 \times 10^{-2})}{1.2 \times 10^{-6}} = \dfrac{(2.4)(4.5)}{1.2} \times \dfrac{10^5 \cdot 10^{-2}}{10^{-6}}$

$= 9.0 \times 10^9$

14. $8x^2 - 4x + 6x + 2 = 8x^2 + 2x + 2$

15. $(5x^2 - 3x + 4) - (2x^2 - 7x - 2)$

$= 5x^2 - 3x + 4 - 2x^2 + 7x + 2$

$= (5x^2 - 2x^2) + (-3x + 7x) + (4 + 2)$

$= 3x^2 + 4x + 6$

16. $(6x - 8) - (3x - 4) = 6x - 8 - 3x + 4$

$= (6x - 3x) + (-8 + 4)$

$= 3x - 4$

17. $2y^2 - 3y - 4$ when $y = -2$

$2(-2)^2 - 3(-2) - 4 = 2 \cdot 4 + 6 - 4$

$= 8 + 6 - 4$

$= 10$

18. $2a^2(3a^2 - 5a + 4) = 2a^2(3a^2) + 2a^2(-5a) + 2a^2(4)$

$= 6a^4 - 10a^3 + 8a^2$

© 2000 Harcourt, Inc

19. $\left(x+\dfrac{1}{2}\right)\left(x+\dfrac{1}{3}\right) = x(x) + x\left(\dfrac{1}{3}\right) + x\left(\dfrac{1}{2}\right) + \left(\dfrac{1}{2}\right)\left(\dfrac{1}{3}\right)$

$= x^2 + \dfrac{1}{3}x + \dfrac{1}{2}x + \dfrac{1}{6}$

$= x^2 + \dfrac{5}{6}x + \dfrac{1}{6} \qquad \dfrac{1}{3}+\dfrac{1}{2}=\dfrac{2}{6}+\dfrac{3}{6}=\dfrac{5}{6}$

20. $(4x-5)(2x+3) = 4x(2x) + 4x(3) + (-5)(2x) + (-5)(3)$

$= 8x^2 + 12x - 10x - 15$

$= 8x^2 + 2x - 15$

21. $\begin{array}{r} x^2 + 3x + 9 \\ x - 3 \\ \hline x^3 + 3x^2 + 9x \\ -3x^2 - 9x - 27 \\ \hline x^3 \qquad\quad - 27 \end{array}$

22. $(x+5)^2 = (x+5)(x+5)$

$= x^2 + 2 \cdot 5x + 5^2$

$= x^2 + 10x + 25$

23. $(3a-2b)^2 = (3a)^2 + 2(3a)(-2b) + (-2b)^2$

$= 9a^2 - 12ab + 4b^2$

24. $(3x-4y)(3x+4y) = (3x)^2 - (4y)^2$

$= 9x^2 - 16y^2$

25. $(a^2-3)(a^2+3) = (a^2)^2 - 3^2$

$= a^4 - 9$

26. $\dfrac{10x^3 + 15x^2 - 5x}{5x} = \dfrac{10x^3}{5x} + \dfrac{15x^2}{5x} - \dfrac{5x}{5x}$

$= 2x^2 + 3x - 1$

27. $\dfrac{8x^2 - 6x - 5}{2x - 3}$

$$\begin{array}{r} 4x + 3 \\ 2x-3 \overline{\smash{)}\, 8x^2 - 6x - 5} \\ -+ \\ \underline{8x^2 - 12x} \\ 6x - 5 \\ -+ \\ \underline{6x - 9} \\ 4 \end{array}$$

The answer is: $4x + 3 + \dfrac{4}{2x-3}$

28. $\dfrac{3x^3 - 2x + 1}{x - 3}$

$$\begin{array}{r} 3x^2 + 9x + 25 \\ x-3 \overline{\smash{)}\, 3x^3 + 0x^2 - 2x + 1} \\ -+ \\ \underline{3x^3 - 9x^2} \\ 9x^2 - 2x \\ -+ \\ \underline{9x^2 - 27x} \\ 25x + 1 \\ -+ \\ \underline{25x - 75} \\ 76 \end{array}$$

The answer is: $3x^2 + 9x + 25 + \dfrac{76}{x-3}$

Chapter 5 Test

29. $V = s^3$
$V = (2.5)^3$
$V = (2.5)(2.5)(2.5)$
$V = 15.625$ centimeters^3

30. x = width
$5x$ = length
$\dfrac{x}{5}$ = height
$V = W \cdot L \cdot H$
$V = x(5x)\left(\dfrac{x}{5}\right)$
$V = x^3$

Chapter 6
Factoring

6.1 The Greatest Common Factor and Factoring by Grouping

1. $15x + 25 = 5(3x + 5)$
3. $6a + 9 = 3(2a + 3)$
5. $4x - 8y = 4(x - 2y)$
7. $3x^2 - 6x - 9 = 3(x^2 - 2x - 3)$
9. $3a^2 - 3a - 60 = 3(a^2 - a - 20)$
11. $24y^2 - 52y + 24 = 4(6y^2 - 13y + 6)$
13. $9x^2 - 8x^3 = x^2(9 - 8x)$
15. $13a^2 - 26a^3 = 13a^2(1 - 2a)$
17. $21x^2y - 28xy^2 = 7xy(3x - 4y)$
19. $22a^2b^2 - 11ab^2 = 11ab^2(2a - 1)$
21. $7x^3 + 21x^2 - 28x = 7x(x^2 + 3x - 4)$
23. $121y^4 - 11x^4 = 11(11y^4 - x^4)$
25. $100x^4 - 50x^3 + 25x^2 = 25x^2(4x^2 - 2x + 1)$
27. $8a^2 + 16b^2 + 32c^2 = 8(a^2 + 2b^2 + 4c^2)$
29. $4a^2b - 16ab^2 + 32a^2b^2 = 4ab(a - 4b + 8ab)$
31. $121a^3b^2 - 22a^2b^3 + 33a^3b^3 = 11a^2b^2(11a - 2b + 3ab)$
33. $12x^2y^3 - 72x^5y^3 - 36x^4y^4 = 12x^2y^3(1 - 6x^3 - 3x^2y)$
35. $xy + 5x + 3y + 15 = x(y + 5) + 3(y + 5) = (y + 5)(x + 3)$
37. $xy + 6x + 2y + 12 = x(y + 6) + 2(y + 6) = (y + 6)(x + 2)$
39. $ab + 7a - 3b - 21 = a(b + 7) - 3(b + 7) = (b + 7)(a - 3)$
41. $ax - bx + ay - by = x(a - b) + y(a - b) = (a - b)(x + y)$
43. $2ax + 6x - 5a - 15 = 2x(a + 3) - 5(a + 3) = (a + 3)(2x - 5)$
45. $3xb - 4b - 6x + 8 = b(3x - 4) - 2(3x - 4) = (3x - 4)(b - 2)$
47. $x^2 + ax + 2x + 2a = x(x + a) + 2(x + a) = (x + a)(x + 2)$
49. $x^2 - ax - bx + ab = x(x - a) - b(x - a) = (x - a)(x - b)$
51. $ax + ay + bx + by + cx + cy = a(x + y) + b(x + y) + c(x + y) = (x + y)(a + b + c)$
53. $6x^2 + 9x + 4x + 6 = 3x(2x + 3) + 2(2x + 3) = (2x + 3)(3x + 2)$
55. $20x^2 - 2x + 50x - 5 = 2x(10x - 1) + 5(10x - 1) = (10x - 1)(2x + 5)$
57. $20x^2 + 4x + 25x + 5 = 4x(5x + 1) + 5(5x + 1) = (5x + 1)(4x + 5)$
59. $x^3 + 2x^2 + 3x + 6 = x^2(x + 2) + 3(x + 2) = (x + 2)(x^2 + 3)$
61. $6x^3 - 4x^2 + 15x - 10 = 2x^2(3x - 2) + 5(3x - 2) = (3x - 2)(2x^2 + 5)$
63. Its greatest common factor is $3 \cdot 2 = 6$.
65. The correct factoring is: $12x^2 + 6x + 3 = 3(4x^2 + 2x + 1)$
67. $A = 1000 + 1000r = 1000(1 + r)$
69. $A = 1,000,000 + 1,000,000r$
    (a) $A = 1,000,000(1 + r)$
    (b) $A = 1,000,000(1 + 0.3) = 1,000,000(1.3)$
        $= 1,300,000$ bacteria

Problem Set 6.2

71. $A = 7{,}000 - 7{,}000r$
    (a) $A = 7{,}000(1-r)$
    (b) $A = 7{,}000(1-0.23)$
    $= 7{,}000(0.77)$
    $= 5390$ names

73. $(x+7)(x-2) = x^2 + 7x - 2x - 14 = x^2 + 5x - 14$

75. $(x-3)(x+2) = x^2 - 3x + 2x - 6 = x^2 - x - 6$

77. $$\begin{array}{r} x^2 - 3x + 9 \\ x+3 \\ \hline x^3 - 3x^2 + 9x \\ 3x^2 - 9x + 27 \\ \hline x^3 + 27 \end{array}$$

79. $$\begin{array}{r} x^2 + 4x - 3 \\ 2x+1 \\ \hline 2x^3 + 8x^2 - 6x \\ x^2 + 4x - 3 \\ \hline 2x^3 + 9x^2 - 2x - 3 \end{array}$$

6.2 Factoring Trinomials

1. $x^2 + 7x + 12 = (x+3)(x+4)$
3. $x^2 + 3x + 2 = (x+2)(x+1)$
5. $a^2 + 10a + 21 = (a+7)(a+3)$
7. $x^2 - 7x + 10 = (x-5)(x-2)$
9. $y^2 - 10y + 21 = (y-7)(y-3)$
11. $x^2 - x - 12 = (x-4)(x+3)$
13. $y^2 + y - 12 = (y+4)(y-3)$
15. $x^2 + 5x - 14 = (x+7)(x-2)$
17. $r^2 - 8r - 9 = (r-9)(r+1)$
19. $x^2 - x - 30 = (x-6)(x+5)$
21. $a^2 + 15a + 56 = (a+7)(a+8)$
23. $y^2 - y - 42 = (y-7)(y+6)$
25. $x^2 + 13x + 42 = (x+7)(x+6)$
27. $2x^2 + 6x + 4 = 2(x^2 + 3x + 2) = 2(x+2)(x+1)$
29. $3a^2 - 3a - 60 = 3(a^2 - a - 20) = 3(a-5)(a+4)$
31. $100x^2 - 500x + 600 = 100(x^2 - 5x + 6) = 100(x-3)(x-2)$
33. $100p^2 - 1300p + 4000 = 100(p^2 - 13p + 40) = 100(p-5)(p-8)$
35. $x^4 - x^3 - 12x^2 = x^2(x^2 - x - 12) = x^2(x-4)(x+3)$
37. $2r^3 + 4r^2 - 30r = 2r(r^2 + 2r - 15) = 2r(r+5)(r-3)$
39. $2y^4 - 6y^3 - 8y^2 = 2y^2(y^2 - 3y - 4) = 2y^2(y-4)(y+1)$
41. $x^5 + 4x^4 + 4x^3 = x^3(x^2 + 4x + 4) = x^3(x+2)(x+2) = x^3(x+2)^2$
43. $3y^4 - 12y^3 - 15y^2 = 3y^2(y^2 - 4y - 5) = 3y^2(y-5)(y+1)$
45. $4x^4 - 52x^3 + 144x^2 = 4x^2(x^2 - 13x + 36) = 4x^2(x-9)(x-4)$

© 2000 Harcourt, Inc

47. $x^2 + 5xy + 6y^2 = (x+2y)(x+3y)$

49. $x^2 - 9xy + 20y^2 = (x-4y)(x-5y)$

51. $a^2 + 2ab - 8b^2 = (a+4b)(a-2b)$

53. $a^2 - 10ab + 25b^2 = (a-5b)(a-5b) = (a-5b)^2$

55. $a^2 + 10ab + 25b^2 = (a+5b)(a+5b) = (a+5b)^2$

57. $x^2 + 2xa - 48a^2 = (x+8a)(x-6a)$

59. $x^2 - 5xb - 36b^2 = (x-9b)(x+4b)$

61. $x^4 - 5x^2 + 6 = (x^2 - 2)(x^2 - 3)$

63. $x^2 - 80x - 2000 = (x-100)(x+20)$

65. $x^2 - x - \dfrac{1}{4} = \left(x - \dfrac{1}{2}\right)\left(x - \dfrac{1}{2}\right) = \left(x - \dfrac{1}{2}\right)^2$

67. $x^2 + 0.6x + 0.08 = (x+0.4)(x+0.2)$

69. Use long division to find the other factor:

$$\begin{array}{r} x+16 \\ x+8 \overline{\smash{)}\, x^2 + 24x + 128} \\ \underline{x^2 + 8x} \\ 16x + 128 \\ \underline{16x + 128} \\ 0 \end{array}$$

The other factor is $x + 16$.

71. $(4x+3)(x-1) = 4x^2 + 3x - 4x - 3 = 4x^2 - x - 3$

73. $(6a+1)(a+2) = 6a^2 + a + 12a + 2 = 6a^2 + 13a + 2$

75. $(3a+2)(2a+1) = 3a(2a) + 3a(1) + 2(2a) + 2(1) = 6a^2 + 3a + 4a + 2 = 6a^2 + 7a + 2$

77. $(6a+2)(a+1) = 6a^2 + 6a + 2a + 2 = 6a^2 + 8a + 2$

79. $(5x^2 + 5x - 4) - (3x^2 - 2x + 7) = 5x^2 + 5x - 4 - 3x^2 + 2x - 7 = 2x^2 + 7x - 11$

81. $(7x+3) - (4x-5) = 7x + 3 - 4x + 5 = 3x + 8$

83. $(5x^2 - 5) - (2x^2 - 4x) = 5x^2 - 5 - 2x^2 + 4x = 3x^2 + 4x - 5$

Problem Set 6.3

6.3 More Trinomials to Factor

1. $2x^2 + 7x + 3 = (2x+1)(x+3)$
3. $2a^2 - a - 3 = (2a-3)(a+1)$
5. $3x^2 + 2x - 5 = (3x+5)(x-1)$
7. $3y^2 - 14y - 5 = (3y+1)(y-5)$
9. $6x^2 + 13x + 6 = (3x+2)(2x+3)$
11. $4x^2 - 12xy + 9y^2 = (2x-3y)(2x-3y) = (2x-3y)^2$
13. $4y^2 - 11y - 3 = (4y+1)(y-3)$
15. $20x^2 - 41x + 20 = (4x-5)(5x-4)$
17. $20a^2 + 48ab - 5b^2 = (10a-b)(2a+5b)$
19. $20x^2 - 21x - 5 = (4x-5)(5x+1)$
21. $12m^2 + 16m - 3 = (6m-1)(2m+3)$
23. $20x^2 + 37x + 15 = (4x+5)(5x+3)$
25. $12a^2 - 25ab + 12b^2 = (3a-4b)(4a-3b)$
27. $3x^2 - xy - 14y^2 = (3x-7y)(x+2y)$
29. $14x^2 + 29x - 15 = (2x+5)(7x-3)$
31. $6x^2 - 43x + 55 = (3x-5)(2x-11)$
33. $15t^2 - 67t + 38 = (5t-19)(3t-2)$
35. $4x^2 + 2x - 6 = 2(2x^2 + x - 3) = 2(2x+3)(x-1)$
37. $24a^2 - 50a + 24 = 2(12a^2 - 25a + 12) = 2(4a-3)(3a-4)$
39. $10x^3 - 23x^2 + 12x = x(10x^2 - 23x + 12) = x(5x-4)(2x-3)$
41. $6x^4 - 11x^3 - 10x^2 = x^2(6x^2 - 11x - 10) = x^2(3x+2)(2x-5)$
43. $10a^3 - 6a^2 - 4a = 2a(5a^2 - 3a - 2) = 2a(5a+2)(a-1)$
45. $15x^3 - 102x^2 - 21x = 3x(5x^2 - 34x - 7) = 3x(5x+1)(x-7)$
47. $35y^3 - 60y^2 - 20y = 5y(7y^2 - 12y - 4) = 5y(7y+2)(y-2)$
49. $15a^4 - 2a^3 - a^2 = a^2(15a^2 - 2a - 1) = a^2(5a+1)(3a-1)$
51. $24x^2y - 6xy - 45y = 3y(8x^2 - 2x - 15) = 3y(4x+5)(2x-3)$
53. $12x^2y - 34xy^2 + 14y^3 = 2y(6x^2 - 17xy + 7y^2) = 2y(2x-y)(3x-7y)$

55. When $x = 2$:
$2x^2 + 7x + 3 = 2(2)^2 + 7(2) + 3 = 8 + 14 + 3 = 25$
$(2x+1)(x+3) = (2 \cdot 2 + 1)(2+3) = (5)(5) = 25$

57. $(2x+3)(2x-3) = (2x)^2 - (3)^2 = 4x^2 - 9$

59. $(x+3)(x-3)(x^2+9) = (x^2-9)(x^2+9) = (x^2)^2 - (9)^2 = x^4 - 81$

61. $h = 8 + 62t - 16t^2$
$= 2(4 + 31t - 8t^2)$
$= 2(4-t)(1+8t)$

See the table in the back of the textbook.

63. $V = x(99 - 40x + 4x^2)$
    $= x(9 - 2x)(11 - 2x)$
    Dimensions of box = dimensions of cardboard $-2x$; $11-2x$ and $9-2x$
    Dimensions of cardboard: 11 inches by 9 inches.

65. $(x+3)(x-3) = x^2 - (3)^2 = x^2 - 9$

67. $(6a+1)(6a-1) = (6a)^2 - (1)^2 = 36a^2 - 1$

69. $(x+4)^2 = x^2 + 2(x)(4) + (4)^2 = x^2 + 8x + 16$

71. $(2x+3)^2 = (2x)^2 + 2(2x)(3) + (3)^2 = 4x^2 + 12x + 9$

6.4 The Difference of Two Squares

1. $x^2 - 9 = (x+3)(x-3)$

3. $a^2 - 36 = (a+6)(a-6)$

5. $x^2 - 49 = (x+7)(x-7)$

7. $4a^2 - 16 = 4(a+2)(a-2)$

9. $9x^2 + 25$ Cannot be factored

11. $25x^2 - 169 = (5x+13)(5x-13)$

13. $9a^2 - 16b^2 = (3a+4b)(3a-4b)$

15. $9 - m^2 = (3+m)(3-m)$

17. $25 - 4x^2 = (5+2x)(5-2x)$

19. $2x^2 - 18 = 2(x^2 - 9) = 2(x+3)(x-3)$

21. $32a^2 - 128 = 32(a^2 - 4) = 32(a+2)(a-2)$

23. $8x^2y - 18y = 2y(4x^2 - 9) = 2y(2x+3)(2x-3)$

25. $a^4 - b^4 = (a^2 + b^2)(a^2 - b^2) = (a^2 + b^2)(a+b)(a-b)$

27. $16m^4 - 81 = (4m^2 + 9)(4m^2 - 9) = (4m^2 + 9)(2m+3)(2m-3)$

29. $3x^3y - 75xy^3 = 3xy(x^2 - 25y^2) = 3xy(x+5y)(x-5y)$

31. $x^2 - 2x + 1 = (x-1)(x-1) = (x-1)^2$

33. $x^2 + 2x + 1 = (x+1)(x+1) = (x+1)^2$

35. $a^2 - 10a + 25 = (a-5)(a-5) = (a-5)^2$

37. $y^2 + 4y + 4 = (y+2)(y+2) = (y+2)^2$

39. $x^2 - 4x + 4 = (x-2)(x-2) = (x-2)^2$

41. $m^2 - 12m + 36 = (m-6)(m-6) = (m-6)^2$

43. $4a^2 + 12a + 9 = (2a+3)(2a+3) = (2a+3)^2$

45. $49x^2 - 14x + 1 = (7x-1)(7x-1) = (7x-1)^2$

47. $9y^2 - 30y + 25 = (3y-5)(3y-5) = (3y-5)^2$

49. $x^2 + 10xy + 25y^2 = (x+5y)(x+5y) = (x+5y)^2$

Problem Set 6.4

51. $9a^2 + 6ab + b^2 = (3a+b)(3a+b) = (3a+b)^2$

53. $3a^2 + 18a + 27 = 3(a^2 + 6a + 9) = 3(a+3)(a+3) = 3(a+3)^2$

55. $2x^2 + 20xy + 50y^2 = 2(x^2 + 10xy + 25y^2) = 2(x+5y)(x+5y) = 2(x+5y)^2$

57. $5x^3 + 30x^2y + 45xy^2 = 5x(x^2 + 6xy + 9y^2) = 5x(x+3y)(x+3y) = 5x(x+3y)^2$

59. $x^2 + 6x + 9 - y^2 = (x+3)^2 - y^2 = (x+3+y)(x+3-y)$

61. $x^2 + 2xy + y^2 - 9 = (x+y)^2 - 9 = (x+y+3)(x+y-3)$

63. Since $(x+7)^2 = x^2 + 14x + 49$, the value is $b = 14$.

65. Since $(x+5)^2 = x^2 + 10x + 25$, the value is $c = 25$.

67. (a) $A = x^2 - 4^2 = x^2 - 16$
    (b) $A = (x+4)(x-4)$
    (c) See the figure in the back of the textbook.

69. $A = a^2 - b^2 = (a+b)(a-b)$

71. $\begin{array}{r} x-2 \\ x-3 \overline{\smash{)}x^2 - 5x + 8} \\ \underline{x^2 - 3x} \\ -2x + 8 \\ \underline{-2x + 6} \\ 2 \end{array}$

$\dfrac{x^2 - 5x + 8}{x-3} = x - 2 + \dfrac{2}{x-3}$

73. $\begin{array}{r} 3x-2 \\ 2x+3 \overline{\smash{)}6x^2 + 5x + 3} \\ \underline{6x^2 + 9x} \\ -4x + 3 \\ \underline{-4x - 6} \\ 9 \end{array}$

$3x - 2 + \dfrac{9}{2x+3}$

6.5 Factoring: A General Review

1. $x^2 - 81 = (x+9)(x-9)$

3. $x^2 + 2x - 15 = (x+5)(x-3)$

5. $x^2 + 6x + 9 = (x+3)(x+3) = (x+3)^2$

7. $y^2 - 10y + 25 = (y-5)(y-5) = (y-5)^2$

9. $2a^3b + 6a^2b + 2ab = 2ab(a^2 + 3a + 1)$

11. $x^2 + x + 1$ Cannot be factored

13. $12a^2 - 75 = 3(4a^2 - 25) = 3(2a+5)(2a-5)$

15. $9x^2 - 12xy + 4y^2 = (3x-2y)(3x-2y) = (3x-2y)^2$

17. $4x^3 + 16xy^2 = 4x(x^2 + 4y^2)$

19. $2y^3 + 20y^2 + 50y = 2y(y^2 + 10y + 25) = 2y(y+5)(y+5) = 2y(y+5)^2$

21. $a^6 + 4a^4b^2 = a^4(a^2 + 4b^2)$

23. $xy + 3x + 4y + 12 = x(y+3) + 4(y+3) = (y+3)(x+4)$

25. $x^4 - 16 = (x^2+4)(x^2-4) = (x^2+4)(x+2)(x-2)$

27. $xy - 5x + 2y - 10 = x(y-5) + 2(y-5) = (y-5)(x+2)$

29. $5a^2 + 10ab + 5b^2 = 5(a^2 + 2ab + b^2) = 5(a+b)(a+b) = 5(a+b)^2$

31. $x^2 + 49$ Cannot be factored

33. $3x^2 + 15xy + 18y^2 = 3(x^2 + 5xy + 6y^2) = 3(x+2y)(x+3y)$

35. $2x^2 + 15x - 38 = (2x+19)(x-2)$

37. $100x^2 - 300x + 200 = 100(x^2 - 3x + 2) = 100(x-2)(x-1)$

39. $x^2 - 64 = (x+8)(x-8)$

41. $x^2 + 3x + ax + 3a = x(x+3) + a(x+3) = (x+3)(x+a)$

43. $49a^7 - 9a^5 = a^5(49a^2 - 9) = a^5(7a+3)(7a-3)$

45. $49x^2 + 9y^2$ Cannot be factored

47. $25a^3 + 20a^2 + 3a = a(25a^2 + 20a + 3) = a(5a+3)(5a+1)$

© 2000 Harcourt, Inc

Problem Set 6.5

49. $xa - xb + ay - by = x(a-b) + y(a-b) = (a-b)(x+y)$

51. $48a^4b - 3a^2b = 3a^2b(16a^2 - 1) = 3a^2b(4a+1)(4a-1)$

53. $20x^4 - 45x^2 = 5x^2(4x^2 - 9) = 5x^2(2x+3)(2x-3)$

55. $3x^2 + 35xy - 82y^2 = (3x + 41y)(x - 2y)$

57. $16x^5 - 44x^4 + 30x^3 = 2x^3(8x^2 - 22x + 15) = 2x^3(2x-3)(4x-5)$

59. $2x^2 + 2ax + 3x + 3a = 2x(x+a) + 3(x+a) = (x+a)(2x+3)$

61. $y^4 - 1 = (y^2 + 1)(y^2 - 1) = (y^2 + 1)(y+1)(y-1)$

63. $12x^4y^2 + 36x^3y^3 + 27x^2y^4 = 3x^2y^2(4x^2 + 12xy + 9y^2) = 3x^2y^2(2x+3y)(2x+3y) = 3x^2y^2(2x+3y)^2$

65. $3x - 6 = 9$
    $3x = 15$
    $x = 5$

67. $2x + 3 = 0$
    $2x = -3$
    $x = -\dfrac{3}{2}$

69. $4x + 3 = 0$
    $4x = -3$
    $x = -\dfrac{3}{4}$

71. $x^8 \cdot x^7 = x^{8+7} = x^{15}$

73. $(3x^3)^2 (2x^4)^3 = 9x^6 \cdot 8x^{12} = 72x^{18}$

75. $57,600 = 5.76 \times 10^4$

© 2000 Harcourt, Inc

6.6 Solving Equations by Factoring

1. Set each factor equal to 0:
$$(x+2)(x-1) = 0$$
$x+2 = 0$ or $x-1 = 0$
$x = -2$ $\qquad x = 1$

3. Set each factor equal to 0:
$a - 4 = 0 \qquad a - 5 = 0$
$a = 4 \qquad\quad a = 5$

5. Set each factor equal to 0:
$$x(x+1)(x-3) = 0$$
$x = 0$ or $x+1 = 0$ or $x-3 = 0$
$x = 0$ $\qquad x = -1 \qquad\quad x = 3$

7. Set each factor equal to 0:
$3x + 2 = 0 \qquad 2x + 3 = 0$
$3x = -2 \qquad\quad x = -3$
$x = -\dfrac{2}{3} \qquad\quad x = -\dfrac{3}{2}$

9. Set each factor equal to 0:
$$m(3m+4)(3m-4) = 0$$
$m = 0$ or $3m+4 = 0$ or $3m-4 = 0$
$m = 0$ $\qquad 3m = -4 \qquad\quad 3m = 4$
$m = 0$ $\qquad m = -\dfrac{4}{3} \qquad\quad m = \dfrac{4}{3}$

11. Set each factor equal to 0:
$2y = 0 \qquad 3y + 1 = 0 \qquad 5y + 3 = 0$
$y = 0 \qquad\quad 3y = -1 \qquad\quad 5y = -3$
$\qquad\qquad y = -\dfrac{1}{3} \qquad\quad y = -\dfrac{3}{5}$

13. Solve by factoring:
$$x^2 + 3x + 2 = 0$$
$(x+2)(x+1) = 0$
$x + 2 = 0$ or $x + 1 = 0$
$x = -2 \qquad\quad x = -1$

15. Solve by factoring:
$$x^2 - 9x + 20 = 0$$
$(x-4)(x-5) = 0$
$x - 4 = 0 \qquad x - 5 = 0$
$x = 4 \qquad\quad x = 5$

17. Solve by factoring:
$$a^2 - 2a - 24 = 0$$
$(a+4)(a-6) = 0$
$a + 4 = 0$ or $a - 6 = 0$
$a = -4 \qquad\quad a = 6$

19. Solve by factoring:
$$100x^2 - 500x + 600 = 0$$
$100(x^2 - 5x + 6) = 0$
$100(x-2)(x-3) = 0$
$x - 2 = 0 \qquad x - 3 = 0$
$x = 2 \qquad\quad x = 3$

21. Solve by factoring:
$$x^2 = -6x - 9$$
$x^2 + 6x + 9 = 0$
$(x+3)(x+3) = 0$
$x + 3 = 0$ or $x + 3 = 0$
$x = -3 \qquad\quad x = -3$

23. Solve by factoring:
$$a^2 - 16 = 0$$
$(a+4)(a-4) = 0$
$a + 4 = 0 \qquad a - 4 = 0$
$a = -4 \qquad\quad a = 4$

© 2000 Harcourt, Inc

Problem Set 6.6

25. Solve by factoring:
$2x^2 + 5x - 12 = 0$
$(x+4)(2x-3) = 0$
$x+4 = 0$ or $2x-3 = 0$
$x = -4$ $\quad\quad 2x = 3$
$x = -4$ $\quad\quad x = \dfrac{3}{2}$

27. Solve by factoring:
$9x^2 + 12x + 4 = 0$
$(3x+2)(3x+2) = 0$
$x = -\dfrac{2}{3}$

29. Solve by factoring:
$a^2 + 25 = 10a$
$a^2 - 10a + 25 = 0$
$(a-5)(a-5) = 0$
$a-5 = 0$ or $a-5 = 0$
$a = 5$ $\quad\quad a = 5$

31. Solve by factoring:
$2x^2 = 3x + 20$
$2x^2 - 3x - 20 = 0$
$(2x+5)(x-4) = 0$
$2x+5 = 0$ $\quad\quad x-4 = 0$
$2x = -5$ $\quad\quad x = 4$
$x = -\dfrac{5}{2}$ $\quad\quad x = 4$

33. Solve by factoring:
$3m^2 = 20 - 7m$
$3m^2 + 7m - 20 = 0$
$(m+4)(3m-5) = 0$
$m+4 = 0$ or $3m-5 = 0$
$m = -4$ $\quad\quad 3m = 5$
$m = -4$ $\quad\quad m = \dfrac{5}{3}$

35. Solve by factoring:
$4x^2 - 49 = 0$
$(2x+7)(2x-7) = 0$
$2x+7 = 0$ $\quad\quad 2x-7 = 0$
$2x = -7$ $\quad\quad 2x = 7$
$x = -\dfrac{7}{2}$ $\quad\quad x = \dfrac{7}{2}$

37. Solve by factoring:
$x^2 + 6x = 0$
$x(x+6) = 0$
$x = 0$ or $x+6 = 0$
$x = 0$ $\quad\quad x = -6$

39. Solve by factoring:
$x^2 - 3x = 0$
$x(x-3) = 0$
$x = 0$ $\quad\quad x-3 = 0$
$x = 0$ $\quad\quad x = 3$

41. Solve by factoring:
$2x^2 = 8x$
$2x^2 - 8x = 0$
$2x(x-4) = 0$
$2x = 0$ or $x-4 = 0$
$x = 0$ $\quad\quad x = 4$

43. Solve by factoring:
$3x^2 = 15x$
$3x^2 - 15x = 0$
$3x(x-5) = 0$
$3x = 0$ $\quad\quad x-5 = 0$
$x = 0$ $\quad\quad x = 5$

© 2000 Harcourt, Inc

45. Solve by factoring:
$$1400 = 400 + 700x - 100x^2$$
$$100x^2 - 700x + 1000 = 0$$
$$100(x^2 - 7x + 10) = 0$$
$$100(x-2)(x-5) = 0$$
$$x - 2 = 0 \quad \text{or} \quad x - 5 = 0$$
$$x = 2 \qquad\qquad x = 5$$

47. Solve by factoring:
$$6x^2 = -5x + 4$$
$$6x^2 + 5x - 4 = 0$$
$$(3x+4)(2x-1) = 0$$
$$3x + 4 = 0 \qquad 2x - 1 = 0$$
$$3x = -4 \qquad\quad 2x = 1$$
$$x = -\frac{4}{3} \qquad\quad x = \frac{1}{2}$$

49. Solve by factoring:
$$x(2x-3) = 20$$
$$2x^2 - 3x = 20$$
$$2x^2 - 3x - 20 = 0$$
$$(2x+5)(x-4) = 0$$
$$2x + 5 = 0 \quad \text{or} \quad x - 4 = 0$$
$$2x = -5 \qquad\qquad x = 4$$
$$x = -\frac{5}{2} \qquad\qquad x = 4$$

51. Solve by factoring:
$$t(t+2) = 80$$
$$t^2 + 2t = 80$$
$$t^2 + 2t - 80 = 0$$
$$(t+10)(t-8) = 0$$
$$t + 10 = 0 \qquad t - 8 = 0$$
$$t = -10 \qquad\quad t = 8$$

53. Solve by factoring:
$$4000 = (1300 - 100p)p$$
$$4000 = 1300p - 100p^2$$
$$100p^2 - 1300p + 4000 = 0$$
$$100(p^2 - 13p + 40) = 0$$
$$100(p-5)(p-8) = 0$$
$$p - 5 = 0 \qquad p - 8 = 0$$
$$p = 5 \qquad\quad p = 8$$

55. Solve by factoring:
$$x(14-x) = 48$$
$$14x - x^2 = 48$$
$$-x^2 + 14x - 48 = 0$$
$$x^2 - 14x + 48 = 0$$
$$(x-6)(x-8) = 0$$
$$x - 6 = 0 \qquad x - 8 = 0$$
$$x = 6 \qquad\quad x = 8$$

57. Solve by factoring:
$$(x+5)^2 = 2x + 9$$
$$x^2 + 10x + 25 = 2x + 9$$
$$x^2 + 8x + 16 = 0$$
$$(x+4)(x+4) = 0$$
$$x + 4 = 0 \qquad x + 4 = 0$$
$$x = -4 \qquad\quad x = -4$$

59. Solve by factoring:
$$(y-6)^2 = y - 4$$
$$y^2 - 12y + 36 = y - 4$$
$$y^2 - 13y + 40 = 0$$
$$(y-5)(y-8) = 0$$
$$y - 5 = 0 \qquad y - 8 = 0$$
$$y = 5 \qquad\quad y = 8$$

Problem Set 6.6

61. Solve by factoring:
$$10^2 = (x+2)^2 + x^2$$
$$100 = x^2 + 4x + 4 + x^2$$
$$2x^2 + 4x - 96 = 0$$
$$2(x^2 + 2x - 48) = 0$$
$$2(x+8)(x-6) = 0$$
$$x + 8 = 0 \qquad x - 6 = 0$$
$$x = -8 \qquad x = 6$$

63. Solve by factoring:
$$2x^3 + 11x^2 + 12x = 0$$
$$x(2x^2 + 11x + 12) = 0$$
$$x(2x+3)(x+4) = 0$$
$$x = 0 \qquad 2x + 3 = 0 \qquad x + 4 = 0$$
$$x = 0 \qquad 2x = -3 \qquad x = -4$$
$$x = 0 \qquad x = -\frac{3}{2} \qquad x = -4$$

65. Solve by factoring:
$$4y^3 - 2y^2 - 30y = 0$$
$$y(4y^2 - 2y - 30) = 0$$
$$y(4y+10)(y-3) = 0$$
$$y = 0 \quad \text{or} \quad 4y + 10 = 0 \quad \text{or} \quad y - 3 = 0$$
$$y = 0 \qquad 4y = -10 \qquad y = 3$$
$$y = 0 \qquad y = -\frac{5}{2} \qquad y = 3$$

67. Solve by factoring:
$$8x^3 + 16x^2 = 10x$$
$$8x^3 + 16x^2 - 10x = 0$$
$$2x(4x^2 + 8x - 5) = 0$$
$$2x(2x-1)(2x+5) = 0$$
$$2x = 0 \qquad 2x - 1 = 0 \qquad 2x + 5 = 0$$
$$x = 0 \qquad 2x = 1 \qquad 2x = -5$$
$$x = 0 \qquad x = \frac{1}{2} \qquad x = -\frac{5}{2}$$

69. Solve by factoring:
$$20a^3 = -18a^2 + 18a$$
$$20a^3 + 18a^2 - 18a = 0$$
$$2a(10a^2 + 9a - 9) = 0$$
$$2a(2a+3)(5a-3) = 0$$
$$2a = 0 \quad \text{or} \quad 2a + 3 = 0 \quad \text{or} \quad 5a - 3 = 0$$
$$a = 0 \qquad 2a = -3 \qquad 5a = 3$$
$$a = 0 \qquad a = -\frac{3}{2} \qquad a = \frac{3}{5}$$

71. Solve by factoring:
$$x^3 + 3x^2 - 4x - 12 = 0$$
$$x^2(x+3) - 4(x+3) = 0$$
$$(x+3)(x^2 - 4) = 0$$
$$(x+3)(x+2)(x-2) = 0$$
$$x + 3 = 0 \qquad x + 2 = 0 \qquad x - 2 = 0$$
$$x = -3 \qquad x = -2 \qquad x = 2$$

73. Solve by factoring:
$$x^3 + x^2 - 16x - 16 = 0$$
$$x^2(x+1) - 16(x+1) = 0$$
$$(x^2 - 16)(x+1) = 0$$
$$(x+4)(x-4)(x+1) = 0$$
$$x + 4 = 0 \quad \text{or} \quad x - 4 = 0 \quad \text{or} \quad x + 1 = 0$$
$$x = -4 \qquad x = 4 \qquad x = -1$$

75. Let $x =$ the cost of the suit
    Let $5x =$ the cost of the bicycle
    $$x + 5x = 90$$
    $$6x = 90$$
    $$x = 15$$
    $$5x = 75$$
    The suit costs $15 and the bicycle costs $75.

77. Let $x =$ the cost of the lot
    Let $4x =$ the cost of the house
    $$x + 4x = 3000$$
    $$5x = 3000$$
    $$x = 600$$
    $$4x = 2400$$
    The lot costs $600 and the house costs $2,400.

79. Simplifying using properties of exponents:
    $$2^{-3} = \frac{1}{2^3} = \frac{1}{8}$$

81. Simplifying using properties of exponents:
    $$\frac{x^5}{x^{-3}} = x^{5-(-3)} = x^8$$

83. Simplifying using properties of exponents:
    $$\frac{(x^2)^3}{(x^{-3})^4} = \frac{x^6}{x^{-12}} = x^{6-(-12)} = x^{6+12} = x^{18}$$

85. $0.0056 = 5.6 \times 10^{-3}$

6.7 Applications

1. Let x and $x + 2 =$ the two integers.
    $$x(x+2) = 80$$
    $$x^2 + 2x = 80$$
    $$x^2 + 2x - 80 = 0$$
    $$(x+10)(x-8) = 0$$
    $$x = -10, 8$$
    $$x + 2 = -8, 10$$
    The two numbers are either -10 and -8, or 8 and 10.

3. Let x and $x + 2 =$ the two integers.
    $$x(x+2) = 99$$
    $$x^2 + 2x = 99$$
    $$x^2 + 2x - 99 = 0$$
    $$(x+11)(x-9) = 0$$
    $$x = -11, 9$$
    $$x + 2 = -9, 11$$
    The two numbers are either -11 and -9, or 9 and 11.

5. Let x and $x + 2 =$ the two integers.
    $$x(x+2) = 5(x+x+2) - 10$$
    $$x^2 + 2x = 5(2x+2) - 10$$
    $$x^2 + 2x = 10x + 10 - 10$$
    $$x^2 + 2x = 10x$$
    $$x^2 - 8x = 0$$
    $$x(x-8) = 0$$
    $$x = 0, 8$$
    $$x + 2 = 2, 10$$
    The two numbers are either 0 and 2, or 8 and 10.

7. Let x and $14 - x =$ the two numbers.
    $$x(14-x) = 48$$
    $$14x - x^2 = 48$$
    $$0 = x^2 - 14x + 48$$
    $$0 = (x-8)(x-6)$$
    $$x = 8, 6$$
    $$14 - x = 6, 8$$
    The two numbers are 6 and 8.

Problem Set 6.7

9. Let x and $5x+2=$ the two numbers.
$$x(5x+2) = 24$$
$$5x^2 + 2x = 24$$
$$5x^2 + 2x - 24 = 0$$
$$(5x+12)(x-2) = 0$$
$$x = -\frac{12}{5}, 2$$
$$5x+2 = -10, 12$$

The two numbers are either $-\frac{12}{5}$ and -10, or 2 and 12.

11. Let x and $4x=$ the two numbers.
$$x(4x) = 4(x+4x)$$
$$4x^2 = 4(5x)$$
$$4x^2 = 20x$$
$$4x^2 - 20x = 0$$
$$4x(x-5) = 0$$
$$x = 0, 5$$
$$4x = 0, 20$$

The two numbers are either 0 and 0, or 5 and 20.

13. Let $w=$ the width
Let $w+1=$ the length
$$w(w+1) = 12$$
$$w^2 + w = 12$$
$$w^2 + w - 12 = 0$$
$$(w+4)(w-3) = 0$$
$$w = 3, w = -4 \text{ is impossible}$$
$$w+1 = 4$$

The width is 3 inches and the length is 4 inches.

15. Let $b=$ the base
Let $2b=$ the height
$$\frac{1}{2}b(2b) = 9$$
$$b^2 = 9$$
$$b^2 - 9 = 0$$
$$(b+3)(b-3) = 0$$
$$b = 3$$
$$b = -3 \text{ is impossible}$$

The base is 3 inches.

17. Let x and $x+2=$ the two legs
$$x^2 + (x+2)^2 = 10^2$$
$$x^2 + x^2 + 4x + 4 = 100$$
$$2x^2 + 4x + 4 = 100$$
$$2x^2 + 4x - 96 = 0$$
$$2(x^2 + 2x - 48) = 0$$
$$2(x+8)(x-6) = 0$$
$$x = 6$$
$$x = -8 \text{ is impossible}$$
$$x+2 = 8$$

The legs are 6 inches and 8 inches.

19. Let $x=$ the longer leg
Let $x+1=$ the hypotenuse
$$5^2 + x^2 = (x+1)^2$$
$$25 + x^2 = x^2 + 2x + 1$$
$$25 = 2x + 1$$
$$24 = 2x$$
$$x = 12$$

The longer leg is 12 meters.

21. Setting $C = \$1,400$:
$$1400 = 400 + 700x - 100x^2$$
$$100x^2 - 700x + 1000 = 0$$
$$100(x^2 - 7x + 10) = 0$$
$$100(x-5)(x-2) = 0$$
$$x = 2, 5$$

The company can manufacture either 200 items or 500 items.

23. Setting $C = \$2,200$:
$$2200 = 600 + 1000x - 100x^2$$
$$100x^2 - 1000x + 1600 = 0$$
$$100(x^2 - 10x + 16) = 0$$
$$100(x-2)(x-8) = 0$$
$$x = 2, 8$$

The company can manufacture either 200 videotapes or 800 videotapes.

© 2000 Harcourt, Inc

25. The revenue is given by:
$$R = xp = (1200 - 100p)p$$
Setting $R = \$3,200$:
$$3200 = (1200 - 100p)p$$
$$3200 = 1200p - 100p^2$$
$$100p^2 - 1200p + 3200 = 0$$
$$100(p^2 - 12p + 32) = 0$$
$$100(p-4)(p-8) = 0$$
$$p = 4, 8$$
The company should sell the ribbons for either $4 or $8.

27. The revenue is given by:
$$R = xp = (1700 - 100p)p$$
Setting $R = \$7,000$:
$$7000 = (1700 - 100p)p$$
$$7000 = 1700p - 100p^2$$
$$100p^2 - 1700p + 7000 = 0$$
$$100(p^2 - 17p + 70) = 0$$
$$100(p-7)(p-10) = 0$$
$$p = 7, 10$$
The calculators should be sold for either $7 or $10.

29. (a) Let $x = $ distance of the base from the wall
Let $2x + 2 = $ point on the wall
Use the Pythagorean Theorem
$$x^2 + (2x+2)^2 = 13^2$$
$$x^2 + 4x^2 + 8x + 4 = 169$$
$$5x^2 + 8x - 165 = 0$$
$$(x-5)(5x+33) = 0$$
$$x = 5 \qquad x = -\frac{33}{5} \text{ is impossible}$$
The distance from the base of the wall.
(b) $2x + 2 = 2(5) + 2 = 12$
The ladder reaches up 12 feet.

31. $h(t) = -16t^2 + 396t + 100$
(a) $0 = -16t^2 + 396t + 100$
$$0 = -4(4t^2 - 99t - 25)$$
$$0 = 4t^2 - 99t - 25$$
$$0 = (4t+1)(t-25)$$
$t = 25$ or $t = -1/4$ (not allowed)
It will reach the ground in 25 seconds.
(b) See the table in the back of the textbook.

33. $(5x^3)^2(2x^6)^3 = 25x^6 \cdot 8x^{18} = 200x^{24}$

35. $\dfrac{x^4}{x^{-3}} = x^{4-(-3)} = x^{4+3} = x^7$

37. $(2 \times 10^{-4})(4 \times 10^5) = 8 \times 10^1 = 80$

39. $20ab^2 - 16ab^2 + 6ab^2 = 10ab^2$

41. $2x^2(3x^2 + 3x - 1) = 2x^2(3x^2) + 2x^2(3x) - 2x^2(1) = 6x^4 + 6x^3 - 2x^2$

43. $(3y - 5)^2 = (3y)^2 - 2(3y)(5) + (5)^2 = 9y^2 - 30y + 25$

45. $(2a^2 + 7)(2a^2 - 7) = (2a^2)^2 - (7)^2 = 4a^4 - 49$

Chapter 6 Review

1. $10x - 20 = 10(x - 2)$

3. $5x - 5y = 5(x - y)$

5. $8x + 4 = 4(2x + 1)$

7. $24y^2 - 40y + 48 = 8(3y^2 - 5y + 6)$

9. $49a^3 - 14b^3 = 7(7a^3 - 2b^3)$

11. $xy + bx + ay + ab = x(y + b) + a(y + b) = (x + a)(y + b)$

13. $2xy + 10x - 3y - 15 = 2x(y + 5) - 3(y + 5) = (y + 5)(2x - 3)$

15. $y^2 + 9y + 14 = (y + 7)(y + 2)$

17. $a^2 - 14a + 48 = (a - 8)(a - 6)$

19. $y^2 + 20y + 99 = (y + 9)(y + 11)$

21. $2x^2 + 13x + 15 = (2x + 3)(x + 5)$

23. $5y^2 + 11y + 6 = (5y + 6)(y + 1)$

25. $6r^2 + 5rt - 6t^2 = (3r - 2t)(2r + 3t)$

27. $n^2 - 81 = (n + 9)(n - 9)$

29. $x^2 + 49$: Prime - Cannot be factored

31. $64a^2 - 121b^2 = (8a + 11b)(8a - 11b)$

33. $y^2 + 20y + 100 = (y + 10)(y + 10) = (y + 10)^2$

35. $64t^2 + 16t + 1 = (8t + 1)(8t + 1) = (8t + 1)^2$

37. $4r^2 - 12rt + 9t^2 = (2r - 3t)(2r - 3t) = (2r - 3t)^2$

39. $2x^2 + 20x + 48 = 2(x^2 + 10x + 24) = 2(x + 4)(x + 6)$

41. $3m^3 - 18m^2 - 21m = 3m(m^2 - 6m - 7) = 3m(m + 1)(m - 7)$

43. $8x^2 + 16x + 6 = 2(4x^2 + 8x + 3) = 2(2x + 1)(2x + 3)$

45. $20m^3 - 34m^2 + 6m = 2m(10m^2 - 17m + 3) = 2m(2m - 3)(5m - 1)$

47. $4x^2 + 40x + 100 = 4(x^2 + 10x + 25) = 4(x + 5)(x + 5) = 4(x + 5)^2$

49. $5x^2 - 45 = 5(x^2 - 9) = 5(x + 3)(x - 3)$

51. $6a^3b + 33a^2b^2 + 15ab^3 = 3ab(2a^2 + 11ab + 5b^2) = 3ab(2a + b)(a + 5b)$

53. $4y^6 + 9y^4 = y^4(4y^2 + 9)$

55. $30a^4b + 35a^3b^2 - 15a^2b^3 = 5a^2b(6a^2 + 7ab - 3b^2) = 5a^2b(3a - b)(2a + 3b)$

57. $(x - 5)(x + 2) = 0$
$x - 5 = 0 \quad x + 2 = 0$
$x = 5 \quad x = -2$

59. $m^2 + 3m = 10$
$m^2 + 3m - 10 = 0$
$(m + 5)(m - 2) = 0$
$m + 5 = 0 \quad m - 2 = 0$
$m = -5 \quad m = 2$

61. $m^2 - 9m = 0$
$m(m-9) = 0$
$m = 0 \quad m - 9 = 0$
$m = 0 \quad m = 9$

63. $9x^4 + 9x^3 = 10x^2$
$9x^4 + 9x^3 - 10x^2 = 0$
$x^2(9x^2 + 9x - 10) = 0$
$x^2(3x+5)(3x-2) = 0$
$x = 0, \dfrac{2}{3}, -\dfrac{5}{3}$

65. Let $x =$ the first integer
Let $x + 1 =$ next consecutive
$x(x+1) = 110$
$x^2 + x = 110$
$x^2 + x - 110 = 0$
$(x+11)(x-10) = 0$
$x = -11, 10$
$x + 1 = -10, 11$
The two numbers are either 10 and 11 or -11 and -10.

67. Let x and $20 - x =$ the two numbers.
$x(20 - x) = 75$
$20x - x^2 = 75$
$0 = x^2 - 20x + 75$
$0 = (x - 15)(x - 5)$
$x = 15, 5$
$20 - x = 5, 15$
The two numbers are 5 and 15.

69. Let $b =$ the base
Let $8b =$ the height
$\dfrac{1}{2}b(8b) = 16$
$4b^2 = 16$
$b^2 - 16 = 0$
$4(b^2 - 4) = 0$
$4(b+2)(b-2) = 0$
$b = 2$

Cumulative Review: Chapters 1-6

Cumulative Review: Chapters 1 - 6

1. $-|-9| = -9$

3. $20 - (-9) = 20 + 9 = 29$

5. $\dfrac{9(-2)}{-2} = \dfrac{-18}{-2} = 9$

7. $\dfrac{-3(4-7)-5(7-2)}{-5-2-1} = \dfrac{-3(-3)-5(5)}{-8} = \dfrac{9-25}{-8} = \dfrac{-16}{-8} = 2$

9. $6 - 2(4a+2) - 5 = 6 - 8a - 4 - 5 = -8a - 3$

11. $(9xy)^0 = 1$

13. $\dfrac{50x^8 y^8}{25x^4 y^2} + \dfrac{28x^7 y^7}{14x^3 y} = 2x^4 y^6 + 2x^4 y^6 = 4x^4 y^6$

15. $3x = -18$
$\dfrac{1}{3}(3x) = \dfrac{1}{3}(-18)$
$x = -6$

17. $-\dfrac{x}{3} = 7$
$-3\left(-\dfrac{x}{3}\right) = -3(7)$
$x = -21$

19. Setting each factor equal to 0:
$4m = 0 \quad m - 7 = 0 \quad 2m - 7 = 0$
$m = 0 \quad\quad m = 7 \quad\quad 2m = 7$
$\quad\quad\quad\quad\quad\quad\quad\quad\quad\quad m = \dfrac{7}{2}$

The solutions are 0, 7, and $\dfrac{7}{2}$.

21. $-2x > -8$
$-\dfrac{1}{2}(-2x) < -\dfrac{1}{2}(-8)$
$x < 4$

23. See the graph in the back of the textbook.

25. Checking the point $(0, 0)$: $2(0) + 3(0) = 0 \geq 6$ (false)
See the graph in the back of the textbook.

27. $m = \dfrac{-4-3}{-2-7} = \dfrac{-7}{-9} = \dfrac{7}{9}$

29. $y = -\dfrac{2}{5}x - \dfrac{2}{3}$.

31. The intersection point is $(-2, 1)$.
See the graph in the back of the textbook.

© 2000 Harcourt, Inc

33. Multiplying the first equation by -3
and the second equation by 7:
$$-15x - 21y = 54$$
$$56x + 21y = 28$$
Adding the two equations:
$$41x = 82$$
$$x = 2$$
Substituting into the second equation
$$8(2) + 3y = 4$$
$$16 + 3y = 4$$
$$3y = -12$$
$$y = -4$$
The solution is $(2, -4)$.

35. Multiplying the first equation by -0.04:
$$-0.04x - 0.04y = -200$$
$$0.04x + 0.06y = 270$$
Adding the two equations:
$$0.02y = 70$$
$$y = 3500$$
Substituting into the first equation:
$$x + 3500 = 5000$$
$$x = 1500$$
The solution is $(1500, 3500)$.

37. $n^2 - 5n - 36 = (n-9)(n+4)$

39. $16 - a^2 = (4+a)(4-a)$

41. $45x^2y - 30xy^2 + 5y^3 = 5y(9x^2 - 6xy + y^2) = 5y(3x-y)(3x-y) = 5y(3x-y)^2$

43. $3xy + 15x - 2y - 10 = 3x(y+5) - 2(y+5) = (y+5)(3x-2)$

45. Commutative property of addition

47. $\dfrac{28x^4y^4 - 14x^2y^3 + 21xy^2}{-7xy^2} = \dfrac{28x^4y^4}{-7xy^2} + \dfrac{-14x^2y^3}{-7xy^2} + \dfrac{21xy^2}{-7xy^2} = -4x^3y^2 + 2xy - 3$

49. Let x and $x + 4 = $ the length of each piece
$$x + x + 4 = 72$$
$$2x + 4 = 72$$
$$2x = 68$$
$$x = 34$$
$$x + 4 = 38$$
The pieces are 34 inches and 38 inches in length.

Chapter 6 Test

1. $5x - 10 = 5(x - 2)$

2. $18x^2y - 9xy - 36xy^2 = 9xy(2x - 1 - 4y)$

3. $x^2 + 2ax - 3bx - 6ab = (x^2 + 2ax) + (-3bx - 6ab)$
   $= x(x + 2a) - 3b(x + 2a)$
   $= (x + 2a)(x - 3b)$

4. $xy + 4x - 7y - 28 = (xy + 4x) + (-7y - 28)$
   $= x(y + 4) - 7(y + 4)$
   $= (x - 7)(y + 4)$

5. $x^2 - 5x + 6 = (x - 2)(x - 3)$

6. $x^2 - x - 6 = (x - 3)(x + 2)$

7. $a^2 - 16 = (a - 4)(a + 4)$

8. $x^2 + 25$ Cannot be factored

9. $x^4 - 81 = (x^2 + 9)(x^2 - 9) = (x^2 + 9)(x - 3)(x + 3)$

10. $27x^2 - 75y^2 = 3(9x^2 - 25y^2) = 3(3x - 5y)(3x + 5y)$

11. $x^3 + 5x^2 - 9x - 45 = x^2(x + 5) - 9(x + 5) = (x + 5)(x^2 - 9) = (x + 5)(x + 3)(x - 3)$

12. $x^2 - bx + 5x - 5b = x(x - b) + 5(x - b) = (x - b)(x + 5)$

13. $4a^2 + 22a + 10 = 2(2a^2 + 11a + 5) = 2(2a + 1)(a + 5)$

14. $3m^2 - 3m - 18 = 3(m^2 - m - 6) = 3(m - 3)(m + 2)$

15. $6y^2 + 7y - 5 = (2y - 1)(3y + 5)$

16. $12x^3 - 14x^2 - 10x = 2x(6x^2 - 7x - 5) = 2x(2x + 1)(3x - 5)$

17. Solve by factoring:
    $x^2 + 7x + 12 = 0$
    $(x + 3)(x + 4) = 0$

18. Solve by factoring:
    $x^2 - 4x + 4 = 0$
    $(x - 2)^2 = 0$
    $x - 2 = 0$
    $x = 2$

19. Solve by factoring:
    $x^2 - 36 = 0$
    $(x + 6)(x - 6) = 0$
    $x = -6, 6$

20. Solve by factoring:
    $x^2 = x + 20$
    $x^2 - x - 20 = 0$
    $(x + 4)(x - 5) = 0$
    $x = -4, 5$

21. Solve by factoring:
    $x^2 - 11x = -30$
    $x^2 - 11x + 30 = 0$
    $(x - 6)(x - 5) = 0$
    $x = 5, 6$

22. Solve by factoring:
    $y^3 = 16y$
    $y^3 - 16y = 0$
    $y(y^2 - 16) = 0$
    $y(y + 4)(y - 4) = 0$
    $y = 0, -4, 4$

23. Solve by factoring:
$$2a^2 = a + 15$$
$$2a^2 - a - 15 = 0$$
$$(2a+5)(a-3) = 0$$
$$a = -\frac{5}{2}, 3$$

24. Solve by factoring:
$$30x^3 - 20x^2 = 10x$$
$$30x^3 - 20x^2 - 10x = 0$$
$$10x(3x^2 - 2x - 1) = 0$$
$$10x(3x+1)(x-1) = 0$$
$$x = 0, -\frac{1}{3}, 1$$

25. Let x and $20 - x =$ the two numbers.
$$x(20-x) = 64$$
$$20x - x^2 = 64$$
$$0 = x^2 - 20x + 64$$
$$0 = (x-16)(x-4)$$
$$x = 4, 16$$
$$20 - x = 16, 4$$
The two numbers are 4 and 16.

26. Let x and $x + 2 =$ the two integers.
$$x(x+2) = x + x + 2 + 7$$
$$x^2 + 2x = 2x + 9$$
$$x^2 - 9 = 0$$
$$(x+3)(x-3) = 0$$
$$x = -3, 3$$
$$x + 2 = -1, 5$$
The two integers are either -3 and -1, or 3 and 5.

27. Let $w =$ the width
Let $3w + 5 =$ the length
$$w(3w+5) = 42$$
$$3w^2 + 5w - 42 = 0$$
$$(3w+14)(w-3) = 0$$
$$w = 3$$
$$w = -\frac{14}{3} \text{ is impossible}$$
$$3w + 5 = 14$$

The width is 3 feet and the length is 14 feet.

28. Let x and $2x + 2 =$ the two legs
$$x^2 + (2x+2)^2 = 13^2$$
$$x^2 + 4x^2 + 8x + 4 = 169$$
$$5x^2 + 8x - 165 = 0$$
$$2(x^2 + 2x - 48) = 0$$
$$(5x+33)(x-5) = 0$$
$$x = 5$$
$$x = -\frac{33}{5} \text{ is impossible}$$
$$2x + 2 = 12$$
The two legs are 5 meters and 12 meters in length.

29. Setting $C = \$800$:
$$800 = 200 + 500x - 100x^2$$
$$100x^2 - 500x + 600 = 0$$
$$100(x^2 - 5x + 6) = 0$$
$$100(x-2)(x-3) = 0$$
$$x = 2, 3$$

The company can manufacture either 200 items or 300 items.

30. The revenue is given by:
$$R = xp = (900 - 100p)p$$
Setting $R = \$1,800$:
$$1800 = (900 - 100p)p$$
$$1800 = 900p - 100p^2$$
$$100p^2 - 900p + 1800 = 0$$
$$100(p^2 - 9p + 18) = 0$$
$$100(p-6)(p-3) = 0$$
$$p = 3, 6$$
The manufacturer should sell the items at either \$3 or \$6.

Chapter 7
Rational Expressions

7.1 Reducing Rational Expressions to Lowest Terms

1. $\dfrac{5}{5x-10} = \dfrac{5}{5(x-2)} = \dfrac{1}{x-2}$

3. $\dfrac{a-3}{a^2-9} = \dfrac{1(a-3)}{(a+3)(a-3)} = \dfrac{1}{a+3}$

5. $\dfrac{x+5}{x^2-25} = \dfrac{x+5}{(x+5)(x-5)} = \dfrac{1}{x-5}$

7. $\dfrac{2x^2-8}{4} = \dfrac{2(x^2-4)}{4} = \dfrac{2(x+2)(x-2)}{4} = \dfrac{(x+2)(x-2)}{2}$

9. $\dfrac{2x-10}{3x-6} = \dfrac{2(x-5)}{3(x-2)}$

11. $\dfrac{10a+20}{5a+10} = \dfrac{10(a+2)}{5(a+2)} = \dfrac{2}{1} = 2$

13. $\dfrac{5x^2-5}{4x+4} = \dfrac{5(x+1)(x-1)}{4(x+1)} = \dfrac{5(x-1)}{4}$

15. $\dfrac{x-3}{x^2-6x+9} = \dfrac{1(x-3)}{(x-3)^2} = \dfrac{1}{x-3}$

17. $\dfrac{3x+15}{3x^2+24x+45} = \dfrac{3(x+5)}{3(x+5)(x+3)} = \dfrac{1}{x+3}$

19. $\dfrac{a^2-3a}{a^3+24x+15a} = \dfrac{a(a-3)}{a(a^2-8a+15)}$
$= \dfrac{a(a-3)}{a(a-3)(a-5)} = \dfrac{1}{a-5}$

21. $\dfrac{3x-2}{9x^2-4} = \dfrac{3x-2}{(3x+2)(3x-2)} = \dfrac{1}{3x+2}$

23. $\dfrac{x^2+8x+15}{x^2+5x+6} = \dfrac{(x+5)(x+3)}{(x+2)(x+3)} = \dfrac{x+5}{x+2}$

25. $\dfrac{2m^3-2m^2-12m}{m^2-5m+6} = \dfrac{2m(m^2-m-6)}{m^2-5m+6} = \dfrac{2m(m-3)(m+2)}{(m-2)(m-3)} = \dfrac{2m(m+2)}{m-2}$

27. $\dfrac{x^3+3x^2-4x}{x^3-16x} = \dfrac{x(x^2+3x-4)}{x(x^2-16)} = \dfrac{x(x+4)(x-1)}{x(x+4)(x-4)} = \dfrac{x-1}{x-4}$

29. $\dfrac{4x^3-10x^2+6x}{2x^3+x^2-3x} = \dfrac{2x(2x^2-5x+3)}{x(2x^2+x-3)} = \dfrac{2x(2x-3)(x-1)}{x(2x+3)(x-1)} = \dfrac{2(2x-3)}{2x+3}$

31. $\dfrac{4x^2-12x+9}{4x^2-9} = \dfrac{(2x-3)^2}{(2x+3)(2x-3)} = \dfrac{2x-3}{2x+3}$

33. $\dfrac{x+3}{x^4-81} = \dfrac{x+3}{(x^2+9)(x+3)(x-3)} = \dfrac{1}{(x^2+9)(x-3)}$

35. $\dfrac{3x^2+x-10}{x^4-16} = \dfrac{(3x-5)(x+2)}{(x^2+4)(x^2-4)} = \dfrac{(3x-5)(x+2)}{(x^2+4)(x+2)(x-2)} = \dfrac{3x-5}{(x^2+4)(x-2)}$

© 2000 Harcourt, Inc

37. $\dfrac{42x^3 - 20x^2 - 48x}{6x^2 - 5x - 4} = \dfrac{2x(21x^2 - 10x - 24)}{(2x+1)(3x-4)} = \dfrac{2x(7x+6)(3x-4)}{(2x+1)(3x-4)} = \dfrac{2x(7x+6)}{2x+1}$

39. $\dfrac{xy + 3x + 2y + 6}{xy + 3x + 5y + 15} = \dfrac{x(y+3) + 2(y+3)}{x(y+3) + 5(y+3)} = \dfrac{(y+3)(x+2)}{(y+3)(x+5)} = \dfrac{x+2}{x+5}$

41. $\dfrac{x^2 - 3x + ax - 3a}{x^2 - 3x + bx - 3b} = \dfrac{x(x-3) + a(x-3)}{x(x-3) + b(x-3)} = \dfrac{(x-3)(x+a)}{(x-3)(x+b)} = \dfrac{x+a}{x+b}$

43. $\dfrac{xy + bx + ay + ab}{xy + bx + 3y + 3b} = \dfrac{x(y+b) + a(y+b)}{x(y+b) + 3(y+b)} = \dfrac{(y+b)(x+a)}{(y+b)(x+3)} = \dfrac{x+a}{x+3}$

45. $\dfrac{8}{6} = \dfrac{4}{3}$ 47. $\dfrac{200}{250} = \dfrac{4}{5}$ 49. $\dfrac{32}{4} = \dfrac{8}{1}$

51. See the table in the back of the textbook.

53. The average speed is: $\dfrac{122 \text{ miles}}{3 \text{ hours}} = 40.7$ miles/hour

55. The average speed is: $\dfrac{785 \text{ feet}}{20 \text{ minutes}} = 39.25$ feet/minute

57. The average speed is: $\dfrac{518 \text{ feet}}{40 \text{ seconds}} = 12.95$ feet/second

59. Her average speed on level ground is: $\dfrac{20 \text{ minutes}}{2 \text{ miles}} = 10$ minutes/mile, or $\dfrac{2 \text{ miles}}{20 \text{ minutes}} = 0.1$ miles/minute

    Her average speed downhill is: $\dfrac{40 \text{ minutes}}{6 \text{ miles}} = \dfrac{20}{3}$ minutes/mile, or $\dfrac{6 \text{ miles}}{40 \text{ minutes}} = \dfrac{3}{20}$ miles/minute

61. The average fuel consumption is: $\dfrac{168 \text{ miles}}{3.5 \text{ gallons}} = 48$ miles/gallon

63. Substituting $x = 5$ and $y = 4$: $\dfrac{x^2 - y^2}{x - y} = \dfrac{5^2 - 4^2}{5 - 4} = \dfrac{25 - 16}{5 - 4} = \dfrac{9}{1} = 9$ The result is equal to $5 + 4$.

65. See the table in the back of the textbook.

    The entries are all -1 because the numerator and denominator are opposites.

    $\dfrac{x-3}{3-x} = \dfrac{-1(3-x)}{3-x} = -1$

67. See the table in the back of the textbook.

Problem Set 7.2

69. $\dfrac{27x^5}{9x^2} - \dfrac{45x^8}{15x^5} = 3x^3 - 3x^3 = 0$

71. $\dfrac{72a^3b^7}{9ab^5} = \dfrac{64a^5b^3}{8a^3b} = 8a^2b^2 + 8a^2b^2 = 16a^2b^2$

73. $\dfrac{38x^7 + 42x^5 - 84x^3}{2x^3} = \dfrac{38x^7}{2x^3} + \dfrac{42x^5}{2x^3} - \dfrac{84x^3}{2x^3} = 19x^4 + 21x^2 - 42$

75. $\dfrac{28a^5b^5 + 36ab^4 - 44a^4b}{4ab} = \dfrac{28a^5b^5}{4ab} + \dfrac{36ab^4}{4ab} - \dfrac{44a^4b}{4ab} = 7a^4b^4 + 9b^3 - 11a^3$

7.2 Multiplication and Division of Rational Expressions

1. $\dfrac{x+y}{3} \cdot \dfrac{6}{x+y} = \dfrac{6(x+y)}{3(x+y)} = 2$

3. $\dfrac{2x+10}{x^2} \cdot \dfrac{x^3}{4x+20} = \dfrac{2(x+5)}{x^2} \cdot \dfrac{x^3}{4(x+5)} = \dfrac{x}{2}$

5. $\dfrac{9}{2a-8} \div \dfrac{3}{a-4} = \dfrac{9}{2a-8} \cdot \dfrac{a-4}{3} = \dfrac{9}{2(a-4)} \cdot \dfrac{a-4}{3} = \dfrac{3}{2}$

7. $\dfrac{x+1}{x^2-9} \div \dfrac{2x+2}{x+3} = \dfrac{x+1}{x^2-9} \cdot \dfrac{x+3}{2x+2} = \dfrac{x+1}{(x+3)(x-3)} \cdot \dfrac{x+3}{2(x+1)} = \dfrac{1}{2(x-3)}$

9. $\dfrac{a^2+5a}{7a} \cdot \dfrac{4a^2}{a^2+4a} = \dfrac{a(a+5)}{7a} \cdot \dfrac{4a^2}{a(a+4)} = \dfrac{4a(a+5)}{7(a+4)}$

11. $\dfrac{y^2-5y+6}{2y+4} \div \dfrac{2y-6}{y+2} = \dfrac{y^2-5y+6}{2y+4} \cdot \dfrac{y+2}{2y-6} = \dfrac{(y-2)(y-3)}{2(y+2)} \cdot \dfrac{y+2}{2(y-3)} = \dfrac{y-2}{4}$

13. $\dfrac{2x-8}{x^2-4} \cdot \dfrac{x^2+6x+8}{x-4} = \dfrac{2(x-4)}{(x+2)(x-2)} \cdot \dfrac{(x+2)(x+4)}{(x-4)} = \dfrac{2(x+4)}{x-2}$

15. $\dfrac{x-1}{x^2-x-6} \cdot \dfrac{x^2+5x+6}{x^2-1} = \dfrac{x-1}{(x-3)(x+2)} \cdot \dfrac{(x+2)(x+3)}{(x+1)(x-1)} = \dfrac{x+3}{(x-3)(x+1)}$

17. $\dfrac{a^2+10a+25}{a+5} \div \dfrac{a^2-25}{a-5} = \dfrac{(a+5)(a+5)}{a+5} \cdot \dfrac{a-5}{(a+5)(a-5)} = 1$

19. $\dfrac{y^3-5y^2}{y^4+3y^3+2y^2} \div \dfrac{y^2-5y+6}{y^2-2y-3} = \dfrac{y^3-5y^2}{y^4+3y^3+2y^2} \cdot \dfrac{y^2-2y-3}{y^2-5y+6}$

$= \dfrac{y^2(y-5)}{y^2(y+2)(y+1)} \cdot \dfrac{(y-3)(y+1)}{(y-2)(y-3)}$

$= \dfrac{y-5}{(y+2)(y-2)}$

21. $\dfrac{2x^2+17x+21}{x^2+2x-35} \cdot \dfrac{x^2-25}{2x^2-7x-15} = \dfrac{(2x+3)(x+7)}{(x+7)(x-5)} \cdot \dfrac{(x+5)(x-5)}{(2x+3)(x-5)} = \dfrac{x+5}{x-5}$

23. $\dfrac{2x^2+10x+12}{4x^2+24x+32} \cdot \dfrac{2x^2+18x+40}{x^2+8x+15} = \dfrac{2(x^2+5x+6)}{4(x^2+6x+8)} \cdot \dfrac{2(x^2+9x+20)}{x^2+8x+15}$

$= \dfrac{2(x+2)(x+3)}{4(x+4)(x+2)} \cdot \dfrac{2(x+5)(x+4)}{(x+5)(x+3)}$

$= 1$

25. $\dfrac{2a^2+7a+3}{a^2-16} \div \dfrac{4a^2+8a+3}{2a^2-5a-12} = \dfrac{(2a+1)(a+3)}{(a-4)(a+4)} \cdot \dfrac{(2a+3)(a-4)}{(2a+1)(2a+3)} = \dfrac{a+3}{a+4}$

27. $\dfrac{4y^2-12y+9}{y^2-36} \div \dfrac{2y^2-5y+3}{y^2+5y-6} = \dfrac{4y^2-12y+9}{y^2-36} \cdot \dfrac{y^2+5y-6}{2y^2-5y+3}$

$= \dfrac{(2y-3)^2}{(y+6)(y-6)} \cdot \dfrac{(y+6)(y-1)}{(2y-3)(y-1)}$

$= \dfrac{2y-3}{y-6}$

29. $\dfrac{x^2-1}{6x^2+42x+60} \cdot \dfrac{7x^2+17x+6}{x+1} \cdot \dfrac{6x+30}{7x^2-11x-6} = \dfrac{(x+1)(x-1)}{6(x+2)(x+5)} \cdot \dfrac{(7x+3)(x+2)}{x+1} \cdot \dfrac{6(x+5)}{(7x+3)(x-2)}$

$= \dfrac{x-1}{x-2}$

31. $\dfrac{18x^3+21x^2-60x}{21x^2-25x-4} \cdot \dfrac{28x^2-17x-3}{16x^3+28x^2-30x} = \dfrac{3x(6x^2+7x-20)}{21x^2-25x-4} \cdot \dfrac{28x^2-17x-3}{2x(8x^2+14x-15)}$

$= \dfrac{3x(3x-4)(2x+5)}{(7x+1)(3x-4)} \cdot \dfrac{(7x+1)(4x-3)}{2x(4x-3)(2x+5)}$

$= \dfrac{3}{2}$

33. $(x^2-9)\left(\dfrac{2}{x+3}\right) = \dfrac{(x+3)(x-3)}{1} \cdot \dfrac{2}{x+3} = 2(x-3)$

Problem Set 7.2

35. $a(a+5)(a-5)\left(\dfrac{2}{a^2-25}\right) = \dfrac{a(a+5)(a-5)}{1} \cdot \dfrac{2}{(a+5)(a-5)} = 2a$

37. $(x^2-x-6)\left(\dfrac{x+1}{x-3}\right) = \dfrac{x^2-x-6}{1} \cdot \dfrac{x+1}{x-3} = \dfrac{(x-3)(x+2)(x+1)}{x-3} = (x+2)(x+1)$

39. $(x^2-4x-5)\left(\dfrac{-2x}{x+1}\right) = \dfrac{(x-5)(x+1)}{1} \cdot \dfrac{-2x}{x+1} = -2x(x-5)$

41. $\dfrac{x^2-9}{x^2-3x} \cdot \dfrac{2x+10}{xy+5x+3y+15} = \dfrac{(x+3)(x-3)}{x(x-3)} \cdot \dfrac{2(x+5)}{x(y+5)+3(y+5)} = \dfrac{2(x+3)(-3)(x+5)}{x(x-3)(x+3)(y+5)} = \dfrac{2(x+5)}{x(y+5)}$

43. $\dfrac{2x^2+4x}{x^2-y^2} \cdot \dfrac{x^2+3x+xy+3y}{x^2+5x+6} = \dfrac{2x(x+2)}{(x+y)(x-y)} \cdot \dfrac{x(x+3)+y(x+3)}{(x+2)(x+3)} = \dfrac{2x}{x-y}$

45. $\dfrac{x^3-3x^2+4x-12}{x^4-16} \cdot \dfrac{3x^2+5x-2}{3x^2-10x+3} = \dfrac{x^2(x-3)+4(x-3)}{(x^2+4)(x^2-4)} \cdot \dfrac{(x+2)(3x-1)}{(3x-1)(x-3)}$

$= \dfrac{(x-3)(x^2+4)}{(x^2+4)(x+2)(x-2)} \cdot \dfrac{(x+2)(3x-1)}{(3x-1)(x-3)}$

$= \dfrac{1}{x-2}$

47. $\left(1-\dfrac{1}{2}\right)\left(1-\dfrac{1}{3}\right)\left(1-\dfrac{1}{4}\right)\left(1-\dfrac{1}{5}\right) = \left(\dfrac{2}{2}-\dfrac{1}{2}\right)\left(\dfrac{3}{3}-\dfrac{1}{3}\right)\left(\dfrac{4}{4}-\dfrac{1}{4}\right)\left(\dfrac{5}{5}-\dfrac{1}{5}\right) = \dfrac{1}{2}\cdot\dfrac{2}{3}\cdot\dfrac{3}{4}\cdot\dfrac{4}{5} = \dfrac{1}{5}$

49. $\left(1-\dfrac{1}{2}\right)\left(1-\dfrac{1}{3}\right)\left(1-\dfrac{1}{4}\right)\cdots\left(1-\dfrac{1}{99}\right)\left(1-\dfrac{1}{100}\right) = \left(\dfrac{1}{2}\right)\left(\dfrac{2}{3}\right)\left(\dfrac{3}{4}\right)\cdots\left(\dfrac{98}{99}\right)\left(\dfrac{99}{100}\right) = \dfrac{1}{100}$

51. Since 5,280 feet = 1 mile, the height is: $\dfrac{14,494 \text{ feet}}{5,280 \text{ feet/mile}} \approx 2.7$ miles

53. $\dfrac{1088 \text{ feet}}{1 \text{ second}} \cdot \dfrac{1 \text{ mile}}{5280 \text{ feet}} \cdot \dfrac{60 \text{ seconds}}{1 \text{ minute}} \cdot \dfrac{60 \text{ minutes}}{1 \text{ hour}} \approx 742$ miles/hour

55. $\dfrac{785 \text{ feet}}{20 \text{ minutes}} \cdot \dfrac{60 \text{ minutes}}{1 \text{ hour}} \cdot \dfrac{1 \text{ mile}}{5280 \text{ feet}} \approx 0.45$ miles/hour

57. $\dfrac{518 \text{ feet}}{40 \text{ seconds}} \cdot \dfrac{60 \text{ seconds}}{1 \text{ minute}} \cdot \dfrac{60 \text{ minutes}}{1 \text{ hour}} \cdot \dfrac{1 \text{ mile}}{5280 \text{ feet}} \approx 8.8$ miles/hour

© 2000 Harcourt, Inc

59. Her average speed on level ground is: $\dfrac{2 \text{ miles}}{1/3 \text{ hour}} = 6$ miles/hour

   Her average speed downhill is: $\dfrac{6 \text{ miles}}{2/3 \text{ hour}} = 9$ miles/hour

61. $\dfrac{1}{2} + \dfrac{5}{2} = \dfrac{6}{2} = 3$

63. $2 + \dfrac{3}{4} = \dfrac{2 \cdot 4}{1 \cdot 4} + \dfrac{3}{4} = \dfrac{8}{4} + \dfrac{3}{4} = \dfrac{11}{4}$

65. $\dfrac{1}{10} + \dfrac{3}{14} = \dfrac{1 \cdot 7}{10 \cdot 7} + \dfrac{3 \cdot 5}{14 \cdot 5} = \dfrac{7}{70} + \dfrac{15}{70} = \dfrac{22}{70} = \dfrac{11}{35}$

67. $\dfrac{10x^4}{2x^2} + \dfrac{12x^6}{3x^4} = 5x^2 + 4x^2 = 9x^2$

69. $\dfrac{12a^2b^5}{3ab^3} + \dfrac{14a^4b^7}{7a^3b^5} = 4ab^2 + 2ab^2 = 6ab^2$

7.3 Addition and Subtraction of Rational Expressions

1. $\dfrac{3}{x} + \dfrac{4}{x} = \dfrac{7}{x}$

3. $\dfrac{9}{a} - \dfrac{5}{a} = \dfrac{4}{a}$

5. $\dfrac{1}{x+1} + \dfrac{x}{x+1} = \dfrac{x+1}{x+1} = 1$

7. $\dfrac{y^2}{y-1} - \dfrac{1}{y-1} = \dfrac{y^2-1}{y-1} = \dfrac{(y+1)(y-1)}{y-1} = y+1$

9. $\dfrac{x^2}{x+2} + \dfrac{4x+4}{x+2} = \dfrac{x^2+4x+4}{x+2} = \dfrac{(x+2)^2}{x+2} = x+2$

11. $\dfrac{x^2}{x-2} - \dfrac{4x-4}{x-2} = \dfrac{x^2-4x+4}{x-2} = \dfrac{(x-2)^2}{x-2} = x-2$

13. $\dfrac{x+2}{x+6} - \dfrac{x-4}{x+6} = \dfrac{x+2-x+4}{x+6} = \dfrac{6}{x+6}$

15. $\dfrac{y}{2} - \dfrac{2}{y} = \dfrac{y \cdot y}{2 \cdot y} - \dfrac{2 \cdot 2}{y \cdot 2} = \dfrac{y^2}{2y} - \dfrac{4}{2y} = \dfrac{y^2-4}{2y} = \dfrac{(y+2)(y-2)}{2y}$

17. $\dfrac{1}{2} + \dfrac{a}{3} = \dfrac{1 \cdot 3}{2 \cdot 3} + \dfrac{a \cdot 2}{3 \cdot 2} = \dfrac{3}{6} + \dfrac{2a}{6} = \dfrac{2a+3}{6}$

19. $\dfrac{x}{x+1} + \dfrac{3}{4} = \dfrac{x \cdot 4}{(x+1) \cdot 4} + \dfrac{3 \cdot (x+1)}{4 \cdot (x+1)} = \dfrac{4x}{4(x+1)} + \dfrac{3x+3}{4(x+1)} = \dfrac{4x+3x+3}{4(x+1)} = \dfrac{7x+3}{4(x+1)}$

21. $\dfrac{x+1}{x-2} - \dfrac{4x+7}{5x-10} = \dfrac{x+1}{x-2} - \dfrac{4x+7}{5(x-2)} = \dfrac{5x+5}{5(x-2)} - \dfrac{4x+7}{5(x-2)} = \dfrac{5x+5-4x-7}{5(x-2)} = \dfrac{x-2}{5(x-2)} = \dfrac{1}{5}$

23. $\dfrac{4x-2}{3x+12} - \dfrac{x-2}{x+4} = \dfrac{4x-2}{3(x+4)} - \dfrac{(x-2) \cdot 3}{(x+4) \cdot 3} = \dfrac{4x-2}{3(x+4)} - \dfrac{3x-6}{3(x+4)} = \dfrac{4x-2-3x+6}{3(x+4)} = \dfrac{x+4}{3(x+4)} = \dfrac{1}{3}$

25. $\dfrac{6}{x(x-2)} + \dfrac{3}{x} = \dfrac{6}{x(x-2)} + \dfrac{3(x-2)}{x(x-2)} = \dfrac{6}{x(x-2)} + \dfrac{3x-6}{x(x-2)} = \dfrac{6+3x-6}{x(x-2)} = \dfrac{3x}{x(x-2)} = \dfrac{3}{x-2}$

Problem Set 7.3

27. $\dfrac{4}{a} - \dfrac{12}{a^2+3a} = \dfrac{4(a+3)}{a(a+3)} - \dfrac{12}{a(a+3)} = \dfrac{4a+12}{a(a+3)} - \dfrac{12}{a(a+3)} = \dfrac{4a+12-12}{a(a+3)} = \dfrac{4a}{a(a+3)} = \dfrac{4}{a+3}$

29. $\dfrac{2}{x+5} - \dfrac{10}{x^2-25} = \dfrac{2}{x+5} - \dfrac{10}{(x+5)(x-5)}$

$= \dfrac{2(x-5)}{(x+5)(x-5)} - \dfrac{10}{(x+5)(x-5)}$

$= \dfrac{2x-10-10}{(x+5)(x-5)}$

$= \dfrac{2x-20}{(x+5)(x-5)}$

$= \dfrac{2(x-10)}{(x+5)(x-5)}$

31. $\dfrac{x-4}{x-3} + \dfrac{6}{x^2-9} = \dfrac{(x-4)(x+3)}{(x-3)(x+3)} + \dfrac{6}{(x+3)(x-3)}$

$= \dfrac{x^2-x-12}{(x+3)(x-3)} + \dfrac{6}{(x+3)(x-3)}$

$= \dfrac{x^2-x-12+6}{(x+3)(x-3)}$

$= \dfrac{x^2-x-6}{(x+3)(x-3)}$

$= \dfrac{(x-3)(x+2)}{(x+3)(x-3)}$

$= \dfrac{x+2}{x+3}$

33. $\dfrac{a-4}{a-3} + \dfrac{5}{a^2-a-6} = \dfrac{a-4}{a-3} + \dfrac{5}{(a+2)(a-3)}$

$= \dfrac{a+2}{a+2}\left(\dfrac{a-4}{a-3}\right) + \dfrac{5}{(a+2)(a-3)}$

$= \dfrac{a^2-2a-8}{(a+2)(a-3)} + \dfrac{5}{(a+2)(a-3)}$

$= \dfrac{a^2-2a-3}{(a+2)(a-3)}$

$= \dfrac{(a+1)(a-3)}{(a+2)(a-3)}$

$= \dfrac{a+1}{a+2}$

35. $\dfrac{8}{x^2-16} - \dfrac{7}{x^2-x-12} = \dfrac{8}{(x+4)(x-4)} - \dfrac{7}{(x-4)(x+3)}$

$= \dfrac{8(x+3)}{(x+4)(x-4)(x+3)} - \dfrac{7(x+4)}{(x+4)(x-4)(x+3)}$

$= \dfrac{8x+24}{(x+4)(x-4)(x+3)} - \dfrac{7x+28}{(x+4)(x-4)(x+3)}$

$= \dfrac{8x+24-7x-28}{(x+4)(x-4)(x+3)}$

$= \dfrac{x-4}{(x+4)(x-4)(x+3)}$

$= \dfrac{1}{(x+4)(x+3)}$

© 2000 Harcourt, Inc

37. $\dfrac{4y}{y^2+6y+5} - \dfrac{3y}{y^2+5y+4} = \dfrac{4y}{(y+5)(y+1)} - \dfrac{3y}{(y+4)(y+1)}$

$= \dfrac{4y(y+4)}{(y+5)(y+1)(y+4)} - \dfrac{3y(y+5)}{(y+1)(y+4)(y+5)}$

$= \dfrac{4y^2+16y-3y^2-15y}{(y+5)(y+1)(y+4)}$

$= \dfrac{y^2+y}{(y+5)(y+1)(y+4)}$

$= \dfrac{y(y+1)}{(y+5)(y+1)(y+4)}$

$= \dfrac{y}{(y+5)(y+4)}$

39. $\dfrac{4x+1}{x^2+5x+4} - \dfrac{x+3}{x^2+4x+3} = \dfrac{4x+1}{(x+4)(x+1)} - \dfrac{x+3}{(x+3)(x+1)}$

$= \dfrac{(4x+1)(x+3)}{(x+4)(x+1)(x+3)} - \dfrac{(x+3)(x+4)}{(x+4)(x+1)(x+3)}$

$= \dfrac{4x^2+13x+3}{(x+4)(x+1)(x+3)} - \dfrac{x^2+7x+12}{(x+4)(x+1)(x+3)}$

$= \dfrac{4x^2+13x+3-x^2-7x-12}{(x+4)(x+1)(x+3)}$

$= \dfrac{3x^2+6x-9}{(x+4)(x+1)(x+3)}$

$= \dfrac{3(x+3)(x-1)}{(x+4)(x+1)(x+3)}$

$= \dfrac{3(x-1)}{(x+4)(x+1)}$

41. $\dfrac{1}{x} + \dfrac{x}{3x+9} - \dfrac{3}{x^2+3x} = \dfrac{1}{x} + \dfrac{x}{3(x+3)} - \dfrac{3}{x(x+3)}$

$= \dfrac{3(x+3)}{3(x+3)}\left(\dfrac{1}{x}\right) + \dfrac{x}{x}\left(\dfrac{x}{3(x+3)}\right) - \dfrac{3}{3}\left(\dfrac{3}{x(x+3)}\right)$

$= \dfrac{3x+9}{3x(x+3)} + \dfrac{x^2}{3x(x+3)} - \dfrac{9}{3x(x+3)}$

$= \dfrac{x^2+3x}{3x(x+3)}$

$= \dfrac{x(x+3)}{3x(x+3)}$

$= \dfrac{1}{3}$

Problem Set 7.3

43. See the table in the back of the textbook.

45. See the table in the back of the textbook.

47. $1 + \dfrac{1}{x+2} = \dfrac{1(x+2)}{1(x+2)} + \dfrac{1}{x+2} = \dfrac{x+2}{x+2} + \dfrac{1}{x+2} = \dfrac{x+2+1}{x+2} = \dfrac{x+3}{x+2}$

49. $1 - \dfrac{1}{x+3} = \dfrac{1(x+3)}{1(x+3)} - \dfrac{1}{x+3} = \dfrac{x+3}{x+3} - \dfrac{1}{x+3} = \dfrac{x+3-1}{x+3} = \dfrac{x+2}{x+3}$

51. $x + 2\left(\dfrac{1}{x}\right) = x + \dfrac{2}{x} = \dfrac{x \cdot x}{1 \cdot x} + \dfrac{2}{x} = \dfrac{x^2}{x} + \dfrac{2}{x} = \dfrac{x^2 + 2}{x}$

53. $\dfrac{1}{x} + \dfrac{1}{2x} = \dfrac{2 \cdot 1}{2x} + \dfrac{1}{2x} = \dfrac{3}{2x}$

55. $2x + 3(x - 3) = 6$
$2x + 3x - 9 = 6$
$5x - 9 = 6$
$5x = 15$
$x = 3$

57. $x - 3(x + 3) = x - 3$
$x - 3x - 9 = x - 3$
$-2x - 9 = x - 3$
$-9 = 3x - 3$
$-6 = 3x$
$-2 = x$

59. $7 - 2(3x + 1) = 4x + 3$
$7 - 6x - 2 = 4x + 3$
$-6x + 5 = 4x + 3$
$-10x + 5 = 3$
$-10x = -2$
$x = \dfrac{1}{5}$

61. $x^2 + 5x + 6 = 0$
$(x + 2)(x + 3) = 0$
$x + 2 = 0 \quad \text{or} \quad x + 3 = 0$
$x = -2 \qquad\qquad x = -3$

63. $x^2 - x = 6$
$x^2 - x - 6 = 0$
$(x - 3)(x + 2) = 0$
$x = -2, 3$

65. $x^2 - 5x = 0$
$x(x - 5) = 0$
$x = 0 \quad \text{or} \quad x - 5 = 0$
$x = 0 \qquad\qquad x = 5$

© 2000 Harcourt, Inc

7.4 Equations Involving Rational Expressions

1. $\dfrac{x}{3} + \dfrac{1}{2} = -\dfrac{1}{2}$

   $6\left(\dfrac{x}{3}\right) + 6\left(\dfrac{1}{2}\right) = 6\left(-\dfrac{1}{2}\right)$

   $2x + 3 = -3$

   $2x = -6$

   $x = -3$

   The solution is $x = -3$.

3. $\dfrac{4}{a} = \dfrac{1}{5}$

   $5a\left(\dfrac{4}{a}\right) = 5a\left(\dfrac{1}{5}\right)$

   $20 = a$

   The solution is $a = 20$.

5. $\dfrac{3}{x} + 1 = \dfrac{2}{x}$

   $x\left(\dfrac{3}{x}\right) + x(1) = x\left(\dfrac{2}{x}\right)$

   $3 + x = 2$

   $x = -1$

   The solution is $x = -1$.

7. $\left(\dfrac{3}{a} - \dfrac{2}{a}\right) = \dfrac{1}{5}$

   $5a\left(\dfrac{3}{a} - \dfrac{2}{a}\right) = 5a\left(\dfrac{1}{5}\right)$

   $15 - 10 = a$

   $a = 5$

   The solution is $a = 5$.

9. $\dfrac{3}{x} + 2 = \dfrac{1}{2}$

   $2x\left(\dfrac{3}{x}\right) + 2x(2) = 2x\left(\dfrac{1}{2}\right)$

   $6 + 4x = x$

   $6 = -3x$

   $-2 = x$

   The solution is $x = -2$.

11. $\dfrac{1}{y} - \dfrac{1}{2} = -\dfrac{1}{4}$

    $4y\left(\dfrac{1}{y} - \dfrac{1}{2}\right) = 4y\left(-\dfrac{1}{4}\right)$

    $4 - 2y = -y$

    $4 = y$

    The solution is $y = 4$.

13. $1 - \dfrac{8}{x} = \dfrac{-15}{x^2}$

    $x^2(1) - x^2\left(\dfrac{8}{x}\right) = x^2\left(\dfrac{-15}{x^2}\right)$

    $x^2 - 8x = -15$

    $x^2 - 8x + 15 = 0$

    $(x - 3)(x - 5) = 0$

    $x = 3, 5$

    The solutions are 3 and 5.

15. $\dfrac{x}{2} - \dfrac{4}{x} = -\dfrac{7}{2}$

    $2x\left(\dfrac{x}{2} - \dfrac{4}{x}\right) = 2x\left(-\dfrac{7}{2}\right)$

    $x^2 - 8 = -7x$

    $x^2 + 7x - 8 = 0$

    $(x + 8)(x - 1) = 0$

    $x = -8, 1$

    The solutions are −8, and 1.

Problem Set 7.4

17.
$$\frac{x-3}{2} + \frac{2x}{3} = \frac{5}{6}$$
$$6\left(\frac{x-3}{2}\right) + 6\left(\frac{2x}{3}\right) = 6\left(\frac{5}{6}\right)$$
$$3(x-3) + 2(2x) = 5$$
$$3x - 9 + 4x = 5$$
$$7x - 9 = 5$$
$$7x = 14$$
$$x = \frac{14}{7}$$
$$x = 2$$
The solution is $x = 2$.

19.
$$\frac{x+1}{3} + \frac{x-3}{4} = \frac{1}{6}$$
$$12\left(\frac{x+1}{3} + \frac{x-3}{4}\right) = 12\left(\frac{1}{6}\right)$$
$$4(x+1) + 3(x-3) = 2$$
$$4x + 4 + 3x - 9 = 2$$
$$7x - 5 = 2$$
$$7x = 7$$
$$x = 1$$
The solution is $x = 1$.

21.
$$\frac{6}{x+2} = \frac{3}{5}$$
$$5(x+2)\left(\frac{6}{x+2}\right) = 5(x+2)\left(\frac{3}{5}\right)$$
$$30 = 3(x+2)$$
$$30 = 3x + 6$$
$$24 = 3x$$
$$8 = x$$
The solution is $x = 8$.

23.
$$\frac{3}{y-2} = \frac{2}{y-3}$$
$$(y-2)(y-3) \cdot \frac{3}{y-2} = (y-2)(y-3) \cdot \frac{2}{y-3}$$
$$3(y-3) = 2(y-2)$$
$$3y - 9 = 2y - 4$$
$$y = 5$$
The solution is $y = 5$.

25.
$$\frac{x}{x-2} + \frac{2}{3} = \frac{2}{x-2}$$
$$3(x-2)\left(\frac{x}{x-2}\right) + 3(x-2)\left(\frac{2}{3}\right) = 3(x-2)\left(\frac{2}{x-2}\right)$$
$$3x + 2(x-2) = 6$$
$$3x + 2x - 4 = 6$$
$$5x = 10$$
$$x = 2$$

27.
$$\frac{x}{x-2} + \frac{3}{2} = \frac{9}{2(x-2)}$$
$$2(x-2)\left(\frac{x}{x-2} + \frac{3}{2}\right) = 2(x-2) \cdot \frac{9}{2(x-2)}$$
$$2x + 3(x-2) = 9$$
$$2x + 3x - 6 = 9$$
$$5x - 6 = 9$$
$$5x = 15$$
$$x = 3$$

© 2000 Harcourt, Inc

29.
$$\frac{5}{x+2} + \frac{1}{x+3} = \frac{-1}{x^2+5x+6}$$

$$(x+2)(x+3)\frac{5}{x+2} + (x+2)(x+3)\frac{1}{x+3} = (x+2)(x+3)\frac{-1}{(x+2)(x+3)}$$

$$5(x+3) + x + 2 = -1$$
$$5x + 15 + x + 2 = -1$$
$$6x + 17 = -1$$
$$6x = -18$$
$$x = -3$$

Since $x = -3$ does not check in the original equation, there is no solution.

31.
$$\frac{8}{x^2-4} + \frac{3}{x+2} = \frac{1}{x-2}$$

$$(x+2)(x-2)\left(\frac{8}{x^2-4} + \frac{3}{x+2}\right) = (x+2)(x-2) \cdot \frac{1}{x-2}$$

$$8 + 3(x-2) = 1(x+2)$$
$$8 + 3x - 6 = x + 2$$
$$2x = 0$$
$$x = 0$$

The solution is $x = 0$.

33.
$$\frac{a}{2} + \frac{3}{a-3} = \frac{a}{a-3}$$

$$2(a-3)\frac{a}{2} + 2(a-3)\left(\frac{3}{a-3}\right) = 2(a-3)\frac{a}{a-3}$$

$$a(a-3) + 6 = 2a$$
$$a^2 - 3a + 6 = 2a$$
$$a^2 - 5a + 6 = 0$$
$$(a-2)(a-3) = 0$$
$$a = 2, 3$$

Since $a = 3$ does not check in the original equation, the solution is $a = 2$.

35.
$$\frac{6}{(y+2)(y-2)} = \frac{4}{y(y+2)}$$

$$y(y+2)(y-2) \cdot \frac{6}{(y+2)(y-2)} = y(y+2)(y-2) \cdot \frac{4}{y(y+2)}$$

$$6y = 4(y-2)$$
$$6y = 4y - 8$$
$$2y = -8$$
$$y = -4$$

The solution is $y = -4$.

© 2000 Harcourt, Inc

Problem Set 7.4

37.
$$\frac{2}{a^2-9} = \frac{3}{a^2+a-12}$$
$$\frac{2}{(a+3)(a-3)} = \frac{3}{(a+4)(a-3)}$$
$$(a+3)(a-3)(a+4)\cdot\frac{2}{(a+3)(a-3)} = (a+3)(a-3)(a+4)\cdot\frac{3}{(a+4)(a-3)}$$
$$2(a+4) = 3(a+3)$$
$$2a+8 = 3a+9$$
$$8 = a+9$$
$$-1 = a$$

The solution is $a = -1$.

39.
$$\frac{3x}{x-5} - \frac{2x}{x+1} = \frac{-42}{(x-5)(x+1)}$$
$$(x-5)(x+1)\left(\frac{3x}{x-5} - \frac{2x}{x+1}\right) = (x-5)(x+1)\cdot\frac{-42}{(x-5)(x+1)}$$
$$3x(x+1) - 2x(x-5) = -42$$
$$3x^2 + 3x - 2x^2 + 10x = -42$$
$$x^2 + 13x + 42 = 0$$
$$(x+7)(x+6) = 0$$
$$x = -7,\ -6$$

The solutions are -7 and -6.

41.
$$\frac{2x}{x+2} = \frac{x}{x+3} - \frac{3}{x^2+5x+6}$$
$$(x+2)(x+3)\frac{2x}{x+2} = (x+2)(x+3)\frac{x}{x+3} - (x+2)(x+3)\frac{3}{(x+2)(x+3)}$$
$$2x(x+3) = x(x+2) - 3$$
$$2x^2 + 6x = x^2 + 2x - 3$$
$$x^2 + 4x + 3 = 0$$
$$(x+3)(x+1) = 0$$
$$x = -3,\ -1$$

Since $x = -3$ does not check in the original equation, the solution is $x = -1$.

43.
$$x + \frac{4}{x} = 5$$
$$x(x) + x\left(\frac{4}{x}\right) = x(5)$$
$$x^2 + 4 = 5x$$
$$x^2 - 5x + 4 = 0$$
$$(x-1)(x-4) = 0$$
$$x = 1,\ 4;\ \text{agrees}$$

45. Let x = the number
$$2(x-3) - 5 = 3$$
$$2x - 6 - 5 = 3$$
$$2x - 11 = 3$$
$$2x = 14$$
$$x = 7$$

47. Let w = the width
Let $2w + 5$ = the length
$2w + 2(2w + 5) = 34$
$2w + 4w + 10 = 34$
$6w + 10 = 34$
$6w = 24$
$w = 4$
$2w + 5 = 13$
The length is 13 inches and the width is 4 inches.

49. Let x and $x + 2$ = the two integers
$x(x + 2) = 48$
$x^2 + 2x = 48$
$x^2 + 2x - 48 = 0$
$(x + 8)(x - 6) = 0$
$x = -8, 6$
$x + 2 = -6, 8$
The two integers are either -8 and -6, or 6 and 8.

51. Let x and $x + 2$ = the two legs
$x^2 + (x + 2)^2 = 10^2$
$x^2 + x^2 + 4x + 4 = 100$
$2x^2 + 4x - 96 = 0$
$x^2 + 2x - 48 = 0$
$(x + 8)(x - 6) = 0$
$x = 6$
$x + 2 = 8$
$x = -8$ is impossible
The legs are 6 inches and 8 inches.

7.5 Applications

1. Let x and $3x$ = the two numbers
$\dfrac{1}{x} + \dfrac{1}{3x} = \dfrac{16}{3}$
$3x\left(\dfrac{1}{x}\right) + 3x\left(\dfrac{1}{3x}\right) = 3x\left(\dfrac{16}{3}\right)$
$3 + 1 = 16x$
$4 = 16x$
$\dfrac{4}{16} = x$
$\dfrac{1}{4} = x$
$\dfrac{3}{4} = 3x$
The numbers are 1/4 and 3/4.

3. Let x = the number
$x + \dfrac{1}{x} = \dfrac{13}{6}$
$6x\left(x + \dfrac{1}{x}\right) = 6x\left(\dfrac{13}{6}\right)$
$6x^2 + 6 = 13x$
$6x^2 - 13x + 6 = 0$
$(3x - 2)(2x - 3) = 0$
$x = \dfrac{2}{3}, \dfrac{3}{2}$
The number is 2/3 or 3/2.

Problem Set 7.5

5. Let x = the number

$$\frac{7+x}{9+x} = \frac{5}{7}$$

$$7(9+x)\frac{7+x}{9+x} = 7(9+x)\frac{5}{7}$$

$$7(7+x) = 5(9+x)$$

$$49 + 7x = 45 + 5x$$

$$49 + 2x = 45$$

$$2x = -4$$

$$x = -2$$

The number is -2.

7. Let x and $x+2$ = the two integers

$$\frac{1}{x} + \frac{1}{x+2} = \frac{5}{12}$$

$$12x(x+2)\left(\frac{1}{x} + \frac{1}{x+2}\right) = 12x(x+2)\left(\frac{5}{12}\right)$$

$$12(x+2) + 12x = 5x(x+2)$$

$$12x + 24 + 12x = 5x^2 + 10x$$

$$0 = 5x^2 - 14x - 24$$

$$(5x+6)(x-4) = 0$$

$$x = 4$$

$$x = -\frac{6}{5} \text{ is impossible}$$

$$x + 2 = 6$$

The integers are 4 and 6.

9. Let x = the rate of the boat in still water.

	d	r	t
Upstream	26	$x-3$	$\frac{26}{x-3}$
Downstream	38	$x+3$	$\frac{38}{x+3}$

$$\frac{26}{x-3} = \frac{38}{x+3}$$

$$(x+3)(x-3)\frac{26}{x-3} = (x+3)(x-3)\frac{38}{x+3}$$

$$(x+3)26 = (x-3)38$$

$$26x + 78 = 38x - 114$$

$$78 = 12x - 114$$

$$192 = 12x$$

$$16 = x$$

The speed is the boat in still water is 16 mph.

11. Let x = the plane speed in still air.

	d	r	t
Against Wind	140	$x-20$	$\dfrac{140}{x-20}$
With Wind	160	$x+20$	$\dfrac{160}{x+20}$

$$\frac{140}{x-20} = \frac{160}{x+20}$$
$$(x+20)(x-20) \cdot \frac{140}{x-20} = (x+20)(x-20) \cdot \frac{160}{x+20}$$
$$140(x+20) = 160(x-20)$$
$$140x + 2800 = 160x - 3200$$
$$-20x + 2800 = -3200$$
$$-20x = -6000$$
$$x = 300$$

The plane speed in still air is 300 mph.

13. Let x and $x+20$ = the rates of each plane

	d	r	t
Plane 1	285	$x+20$	$\dfrac{285}{x+20}$
Plane 2	255	x	$\dfrac{255}{x}$

$$\frac{285}{x+20} = \frac{255}{x}$$
$$x(x+20)\frac{285}{x+20} = x(x+20)\frac{255}{x}$$
$$285x = 255(x+20)$$
$$285x = 255x + 5100$$
$$30x = 5100$$
$$x = 170$$
$$x + 20 = 190$$

The plane speeds are 170 mph and 190 mph.

Problem Set 7.5

15. Let x = her rate downhill.

	d	r	t
Level Ground	2	$x-3$	$\dfrac{2}{x-3}$
Downhill	6	x	$\dfrac{6}{x}$

$$\frac{2}{x-3}+\frac{6}{x}=1$$
$$x(x-3)\left(\frac{2}{x-3}+\frac{6}{x}\right)=x(x-3)\cdot 1$$
$$2x+6(x-3)=x(x-3)$$
$$2x+6x-18=x^2-3x$$
$$8x-18=x^2-3x$$
$$0=x^2-11x+18$$
$$0=(x-2)(x-9)$$
$$x=9 \quad (x=2 \text{ is impossible})$$

Tina runs 9 mph on the downhill part of the course.

17. Let x = her rate on level ground.

	d	r	t
Level Ground	4	x	$\dfrac{4}{x}$
Downhill	5	$x+2$	$\dfrac{5}{x+2}$

$$\frac{4}{x}+\frac{5}{x+2}=1$$
$$x(x+2)\left(\frac{4}{x}+\frac{5}{x+2}\right)=x(x+2)\cdot 1$$
$$4(x+2)+5x=x(x+2)$$
$$4x+8+5x=x^2+2x$$
$$9x+8=x^2+2x$$
$$0=x^2-7x-8$$
$$0=(x-8)(x+1)$$
$$x=8 \quad (x=-1 \text{ is impossible})$$

Jerri jogs 8 mph on level ground.

© 2000 Harcourt, Inc

19. Let t = the time to fill the pool with both pipes left open.

$$\frac{1}{12} - \frac{1}{15} = \frac{1}{t}$$

$$60t\left(\frac{1}{12} - \frac{1}{15}\right) = 60t \cdot \frac{1}{t}$$

$$5t - 4t = 60$$

$$t = 60$$

It will take 60 hours to fill the pool with both pipes left open.

21. Let t = the time to fill the bathtub with both faucets open.

$$\frac{1}{10} + \frac{1}{12} = \frac{1}{t}$$

$$60t\left(\frac{1}{10} + \frac{1}{12}\right) = 60t \cdot \frac{1}{t}$$

$$6t + 5t = 60$$

$$11t = 60$$

$$t = \frac{60}{11} = 5\frac{5}{11}$$

It will take 5 5/11 minutes to fill the tub with both faucets open.

23. Let t = the time to fill the sink with both the faucet and the drain left open.

$$\frac{1}{3} - \frac{1}{4} = \frac{1}{t}$$

$$12t\left(\frac{1}{3} - \frac{1}{4}\right) = 12t \cdot \frac{1}{t}$$

$$4t - 3t = 12$$

$$t = 12$$

It will take 12 minutes for the sink to overflow with both the faucet and drain left open.

25. See the graph in the back of the textbook.

27. $y = \dfrac{-4}{x}$

    See the graph in the back of the textbook.

29. $y = \dfrac{8}{x}$

    See the graph in the back of the textbook.

31. See the graph in the back of the textbook.

33. $15a^3b^3 - 20a^2b - 35ab^2 = 5ab(3a^2b^2 - 4a - 7b)$

35. $x^2 - 4x - 12 = (x-6)(x+2)$

37. $x^4 - 16 = (x^2 + 4)(x^2 - 4) = (x^2 + 4)(x+2)(x-2)$

Problem Set 7.6

39. $5x^3 - 25x^2 - 30x = 5x(x^2 - 5x - 6) = 5x(x-6)(x+1)$

41. Solve the equation by factoring:

$$x^2 - 6x = 0$$
$$x(x-6) = 0$$
$$x = 0, 6$$

43. Solving the equation by factoring:

$$x(x+2) = 80$$
$$x^2 + 2x = 80$$
$$x^2 + 2x - 80 = 0$$
$$(x+10)(x-8) = 0$$
$$x = -10, 8$$

45. Let x and $x+3$ = the two legs.

$$x^2 + (x+3)^2 = 15^2$$
$$x^2 + x^2 + 6x + 9 = 225$$
$$2x^2 + 6x - 216 = 0$$
$$x^2 + 3x - 108 = 0$$
$$(x+12)(x-9) = 0$$
$$x = 9 \quad (x = -12 \text{ is impossible})$$
$$x + 3 = 12$$

The two legs are 9 inches and 12 inches.

7.6 Complex Fractions

1. $\dfrac{\frac{3}{4}}{\frac{1}{8}} = \dfrac{8 \cdot \frac{3}{4}}{8 \cdot \frac{1}{8}} = \dfrac{6}{1} = 6$

3. $\dfrac{\frac{2}{3}}{4} = \dfrac{\frac{2}{3} \cdot 3}{4 \cdot 3} = \dfrac{2}{12} = \dfrac{1}{6}$

5. $\dfrac{\frac{x^2}{y}}{\frac{x}{y^3}} = \dfrac{y^3 \cdot \frac{x^2}{y}}{y^3 \cdot \frac{x}{y^3}} = \dfrac{x^2 y^2}{x} = xy^2$

7. $\dfrac{\frac{4x^3}{y^6}}{\frac{8x^2}{y^7}} = \dfrac{\frac{4x^3}{y^6} \cdot y^7}{\frac{8x^2}{y^7} \cdot y^7} = \dfrac{4x^3 y}{8x^2} = \dfrac{xy}{2}$

9. $\dfrac{y+\frac{1}{x}}{x+\frac{1}{y}} = \dfrac{xy\left(y+\frac{1}{x}\right)}{xy\left(x+\frac{1}{y}\right)}$

$= \dfrac{xy^2+y}{x^2y+x}$

$= \dfrac{y(xy+1)}{x(xy+1)}$

$= \dfrac{y}{x}$

11. $\dfrac{1+\frac{1}{a}}{1-\frac{1}{a}} = \dfrac{\left(1+\frac{1}{a}\right)\cdot a}{\left(1-\frac{1}{a}\right)\cdot a} = \dfrac{a+1}{a-1}$

13. $\dfrac{\frac{x+1}{x^2-9}}{\frac{2}{x+3}} = \dfrac{x+1}{x^2-9}\cdot\dfrac{x+3}{2}$

$= \dfrac{x+1}{(x+3)(x-3)}\cdot\dfrac{x+3}{2}$

$= \dfrac{x+1}{2(x-3)}$

15. $\dfrac{\frac{1}{a+2}}{\frac{1}{a^2-a-6}} = \dfrac{\frac{1}{a+2}(a-3)(a+2)}{\frac{1}{(a-3)(a+2)}(a-3)(a+2)}$

$= \dfrac{a-3}{1} = a-3$

17. $\dfrac{1-\frac{9}{y^2}}{1-\frac{1}{y}-\frac{6}{y^2}} = \dfrac{\left(1-\frac{9}{y^2}\right)y^2}{\left(1-\frac{1}{y}-\frac{6}{y^2}\right)y^2} = \dfrac{y^2-9}{y^2-y-6}$

$= \dfrac{(y+3)(y-3)}{(y+2)(y-3)}$

$= \dfrac{y+3}{y+2}$

19. $\dfrac{\frac{1}{y}+\frac{1}{x}}{\frac{1}{xy}} = \dfrac{\left(\frac{1}{y}+\frac{1}{x}\right)xy}{\left(\frac{1}{xy}\right)xy} = \dfrac{x+y}{1} = x+y$

21. $\dfrac{1-\frac{1}{a^2}}{1-\frac{1}{a}} = \dfrac{\left(1-\frac{1}{a^2}\right)a^2}{\left(1-\frac{1}{a}\right)a^2} = \dfrac{a^2-1}{a^2-a} = \dfrac{(a+1)(a-1)}{a(a-1)}$

$= \dfrac{a+1}{a}$

23. $\dfrac{\frac{1}{10x}-\frac{y}{10x^2}}{\frac{1}{10}-\frac{y}{10x}} = \dfrac{\left(\frac{1}{10x}-\frac{y}{10x^2}\right)10x^2}{\left(\frac{1}{10}-\frac{y}{10x}\right)10x^2} = \dfrac{x-y}{x^2-xy}$

$= \dfrac{1(x-y)}{x(x-y)} = \dfrac{1}{x}$

Problem Set 7.6

25. $\dfrac{\dfrac{1}{a+1}+2}{\dfrac{1}{a+1}+3} = \dfrac{\left(\dfrac{1}{a+1}+2\right)(a+1)}{\left(\dfrac{1}{a+1}+3\right)(a+1)} = \dfrac{1+2(a+1)}{1+3(a+1)}$

$= \dfrac{1+2a+2}{1+3a+3} = \dfrac{2a+3}{3a+4}$

27. Simplify each parenthesis first:

$1 - \dfrac{1}{x} = \dfrac{x}{x} - \dfrac{1}{x} = \dfrac{x-1}{x}$

$1 - \dfrac{1}{x+1} = \dfrac{x+1}{x+1} - \dfrac{1}{x+1} = \dfrac{x}{x+1}$

$1 - \dfrac{1}{x+2} = \dfrac{x+2}{x+2} - \dfrac{1}{x+2} = \dfrac{x+1}{x+2}$

Multiply

$\left(1-\dfrac{1}{x}\right)\left(1-\dfrac{1}{x+1}\right)\left(1-\dfrac{1}{x+2}\right)$

$= \dfrac{x-1}{x} \cdot \dfrac{x}{x+1} \cdot \dfrac{x+1}{x+2} = \dfrac{x-1}{x+2}$

29. Simplify each parenthesis first:

$1 + \dfrac{1}{x+3} = \dfrac{x+3}{x+3} + \dfrac{1}{x+3} = \dfrac{x+4}{x+3}$

$1 + \dfrac{1}{x+2} = \dfrac{x+2}{x+2} + \dfrac{1}{x+2} = \dfrac{x+3}{x+2}$

$1 + \dfrac{1}{x+1} = \dfrac{x+1}{x+1} + \dfrac{1}{x+1} = \dfrac{x+2}{x+1}$

Multiply

$\left(1+\dfrac{1}{x+3}\right)\left(1+\dfrac{1}{x+2}\right)\left(1+\dfrac{1}{x+1}\right)$

$= \dfrac{x+4}{x+3} \cdot \dfrac{x+3}{x+2} \cdot \dfrac{x+2}{x+1} = \dfrac{x+4}{x+1}$

31. $2 + \dfrac{1}{2+1} = 2 + \dfrac{1}{3} = \dfrac{6}{3} + \dfrac{1}{3} = \dfrac{7}{3}$

$2 + \dfrac{1}{2+\dfrac{1}{2+1}} = 2 + \dfrac{1}{\dfrac{7}{3}} = 2 + \dfrac{3}{7} = \dfrac{14}{7} + \dfrac{3}{7} = \dfrac{17}{7}$

$2 + \dfrac{2}{2+\dfrac{1}{2+\dfrac{1}{2+1}}} = 2 + \dfrac{1}{\dfrac{17}{7}} = 2 + \dfrac{7}{17} = \dfrac{34}{17} + \dfrac{7}{17} = \dfrac{41}{17}$

33. See the table in the back of the textbook.

35. See the table in the back of the textbook.

37. $2x+3 < 5$
$2x+3-3 < 5-3$
$2x < 2$
$\dfrac{1}{2}(2x) < \dfrac{1}{2}(2)$
$x < 1$

39. $-3x \leq 21$
$-\dfrac{1}{3}(-3x) \geq -\dfrac{1}{3}(21)$
$x \geq -7$

41. $-2x+8 > -4$
$-2x+8-8 > -4-8$
$-2x > -12$
$-\dfrac{1}{2}(-2x) < -\dfrac{1}{2}(-12)$
$x < 6$

43. $4-2(x+1) \geq -2$
$4-2x-2 \geq -2$
$-2x+2 \geq -2$
$-2x+2-2 \geq -2-2$
$-2x \geq -4$
$-\dfrac{1}{2}(-2x) \leq -\dfrac{1}{2}(-4)$
$x \leq 2$

7.7 Proportions

1. $\dfrac{x}{2} = \dfrac{6}{12}$
   $12x = 12$
   $x = 1$

3. $\dfrac{2}{5} = \dfrac{4}{x}$
   $2x = 20$
   $x = 10$

5. $\dfrac{10}{20} = \dfrac{20}{x}$
   $10x = 400$
   $x = 40$

7. $\dfrac{a}{3} = \dfrac{5}{12}$
   $12a = 15$
   $a = \dfrac{15}{12} = \dfrac{5}{4}$

9. $\dfrac{2}{x} = \dfrac{6}{7}$
   $14 = 6x$
   $\dfrac{14}{6} = x$
   $\dfrac{7}{3} = x$

11. $\dfrac{x+1}{3} = \dfrac{4}{x}$
    $x^2 + x = 12$
    $x^2 + x - 12 = 0$
    $(x+4)(x-3) = 0$
    $x = -4, 3$
    The solutions are −4 and 3.

13. $\dfrac{x}{2} = \dfrac{8}{x}$
    $x^2 = 16$
    $x^2 - 16 = 0$
    $(x+4)(x-4) = 0$
    $x = -4, 4$
    The solutions are −4 and 4.

15. $\dfrac{4}{a+2} = \dfrac{a}{2}$
    $a^2 + 2a = 8$
    $a^2 + 2a - 8 = 0$
    $(a+4)(a-2) = 0$
    $a = -4, 2$
    The solutions are −4 and 2.

17. $\dfrac{1}{x} = \dfrac{x-5}{6}$
    $6 = x^2 - 5x$
    $x^2 - 5x - 6 = 0$
    $(x+1)(x-6) = 0$
    $x = -1, 6$

19. Compare hits to games
    $\dfrac{6}{18} = \dfrac{x}{45}$
    $18x = 270$
    $x = 15$
    He will get 15 hits in 45 games.

21. Compare ml alcohol to ml water.
    $\dfrac{12}{16} = \dfrac{x}{28}$
    $16x = 336$
    $x = 21$
    The solution will have 21 ml of alcohol.

23. Compare grams of fat to total grams.
    $\dfrac{13}{100} = \dfrac{x}{350}$
    $100x = 4550$
    $x = 45.5$
    There are 45.5 grams of fat in 350 grams of ice cream.

Problem Set 7.8

25. Compare inches on the map to actual miles.

$$\frac{3.5}{100} = \frac{x}{420}$$
$$100x = 1470$$
$$x = 14.7$$

They are 14.7 inches apart on the map.

27. Compare miles to hours.

$$\frac{245}{5} = \frac{x}{7}$$
$$5x = 1715$$
$$x = 343$$

He will travel 343 miles.

29. $\dfrac{x^2 - x - 6}{x^2 - 9} = \dfrac{(x-3)(x+2)}{(x+3)(x-3)} = \dfrac{x+2}{x+3}$

31. $\dfrac{x^2 - 25}{x+4} \cdot \dfrac{2x+8}{x^2 - 9x + 20} = \dfrac{(x+5)(x-5)}{x+4} \cdot \dfrac{2(x+4)}{(x-5)(x-4)}$

$= \dfrac{2(x+5)(x-5)(x+4)}{(x+4)(x-5)(x-4)} = \dfrac{2(x+5)}{x-4}$

33. $\dfrac{x}{x^2 - 16} + \dfrac{4}{x^2 - 16} = \dfrac{x+4}{x^2 - 16} = \dfrac{1(x+4)}{(x+4)(x-4)} = \dfrac{1}{x-4}$

7.8 Variation

1. $y = kx: y = 10, x = 5$
   $10 = k(5) \Rightarrow k = 2$
   $y = 2x$
   If $x = 4$
   $y = 2(4) = 8$

3. $y = kx: y = 39, x = 3$
   $39 = k(3) \Rightarrow k = 13$
   $k = 13$
   $y = 13x$
   If $x = 10$
   $y = 13(10) = 130$

5. $y = kx: y = -24, x = 4$
   $-24 = k(4) \Rightarrow k = -6$
   $y = -6x$
   If $y = 30$
   $-30 = -6x$
   $5 = x$

7. $y = kx: y = -7, x = -1$
   $-7 = k(-1) \Rightarrow k = 7$
   $y = 7x$
   If $y = -21$
   $-21 = 7x$
   $x = -3$

9. $y = kx^2: y = 75, x = 5$
   $75 = k(5)^2 \Rightarrow k = 3$
   $y = 3x^2$
   If $x = 1$
   $y = 3(1)^2 = 3$

11. $y = kx^2: y = 48, x = 4$
    $48 = k(4)^2 \Rightarrow k = 3$
    $y = 3x^2$
    If $x = 9$
    $y = 3(9)^2 = 243$

© 2000 Harcourt, Inc

13. $y = \dfrac{k}{x}: y = 5, x = 2$

　　$5 = \dfrac{k}{2} \Rightarrow k = 10$

　　$y = \dfrac{10}{x}$

　　If $x = 5$

　　$y = \dfrac{10}{5} = 2$

15. $y = \dfrac{k}{x}: y = 2, x = 1$

　　$2 = \dfrac{k}{1} \Rightarrow k = 2$

　　$y = \dfrac{2}{x}$

　　If $x = 4$

　　$y = \dfrac{2}{4} = \dfrac{1}{2}$

17. $y = \dfrac{k}{x}: y = 5, x = 3$

　　$5 = \dfrac{k}{3} \Rightarrow k = 15$

　　$y = \dfrac{15}{x}$

　　If $y = 15$

　　$15 = \dfrac{15}{x}$

　　$x = 1$

19. $y = \dfrac{k}{x}: y = 10, x = 10$

　　$10 = \dfrac{k}{10} \Rightarrow k = 100$

　　$y = \dfrac{100}{x}$

　　If $y = 20$

　　$20 = \dfrac{100}{x}$

　　$x = 5$

21. $y = \dfrac{k}{x^2}: y = 4, x = 5$

　　$4 = \dfrac{k}{(5)^2} \Rightarrow k = 100$

　　$y = \dfrac{100}{x^2}$

　　If $x = 2$

　　$y = \dfrac{100}{(2)^2} = 25$

23. $y = \dfrac{k}{x^2}: y = 4, x = 3$

　　$4 = \dfrac{k}{(3)^2} \Rightarrow k = 36$

　　$y = \dfrac{36}{x^2}$

　　If $x = 2$

　　$y = \dfrac{36}{(2)^2} = 9$

25. $t = kd: t = 42, d = 2$

　　$42 = k(2) \Rightarrow k = 21$

　　$t = 21d$

　　If $d = 4$

　　$t = 21(4) = 84$

　　The tension is 84 lbs.

27. $P = kI^2: P = 30, I = 2$

　　$30 = k(2)^2 \Rightarrow k = \dfrac{15}{2}$

　　$P = (15/2)I^2$

　　If $I = 7$

　　$P = \left(\dfrac{15}{2}\right)(7)^2 = 367.5$

Problem Set 7.8

29. $M = kh$: $M = 157$, $h = 20$
$157 = k(20) \Rightarrow k = 7.85$
$M = 7.85h$
If $h = 30$
$M = 7.85(30) = 235.50$

She makes $235.50.

31. $F = \dfrac{k}{d^2}$: $F = 150$, $d = 4000$
$150 = \dfrac{k}{4000^2} \Rightarrow k = 2,400,000,000$
$F = 2,400,000,000 / d^2$
If $d = 5000$
$F = 2,400,000,000 / 5000^2 = 96$ pounds

33. $I = \dfrac{k}{R}$: $I = 30$, $R = 2$
$30 = \dfrac{k}{2} \Rightarrow k = 60$
$I = \dfrac{60}{R}$
If $R = 5$
$I = \dfrac{60}{5} = 12$

35. $2x + y = 3$ (1)
$\underline{3x - y = 7}$ (2)
$5x = 10$
$x = 2$
Substitute 2 for x in (1)
$2(2) + y = 3$
$y = -1$
The solution is $(2, -1)$.

37. $4x - 5y = 1$ (1)
$x - 2y = -2$ (2)
Add -4 times (2) to (1)
$3y = 9$
$y = 3$
Substitute 3 for y in (2)
$x - 2(3) = -2$
$x = 4$
The solution is $(4, 3)$.

39. $5x + 2y = 7$ (1)
$y = 3x - 2$ (2)
Substitute $3x - 2$ for y in (1)
$5x + 2(3x - 2) = 7$
$5x + 6x - 4 = 7$
$11x - 4 = 7$
$11x = 11$
$x = 1$
Substitute 1 for x in (2)
$y = 3(1) - 2$
$y = 3 - 2$
$y = 1$
The solution is $(1, 1)$.

41. $2x - 3y = 4$ (1)
$x = 2y + 1$ (2)
Substitute $2y + 1$ for x in (1)
$2(2y + 1) - 3y = 4$
$4y + 2 - 3y = 4$
$y + 2 = 4$
$y = 2$
Substitute 2 for y in (2)
$x = 2(2) + 1 = 5$
The solution is $(5, 2)$.

Chapter 7 Review

1. $\dfrac{7}{14x-28} = \dfrac{7}{14(x-2)} = \dfrac{1}{2(x-2)} \quad x \neq 2$

3. $\dfrac{8x-4}{4x+12} = \dfrac{4(2x-1)}{4(x+3)} = \dfrac{2x-1}{x+3} \quad x \neq -3$

5. $\dfrac{3x^3+16x^2-12x}{2x^3+9x^2-18x} = \dfrac{x(3x^2+16x-12)}{x(2x^2+9x-18)}$
$= \dfrac{x(x+6)(3x-2)}{x(x+6)(2x-3)}$
$= \dfrac{\cancel{x(x+6)}(3x-2)}{\cancel{x(x+6)}(2x-3)}$
$= \dfrac{3x-2}{2x-3}$

7. $\dfrac{x^2+5x-14}{x+7} = \dfrac{(x+7)(x-2)}{x+7}$
$= \dfrac{\cancel{(x+7)}(x-2)}{\cancel{x+7}}$
$= x-2$

9. $\dfrac{xy+bx+ay+ab}{xy+5x+ay+5a} = \dfrac{(xy+bx)+(ay+ab)}{(xy+5x)+(ay+5a)}$
$= \dfrac{x(y+b)+a(y+b)}{x(y+5)+a(y+5)}$
$= \dfrac{(y+b)(x+a)}{(y+5)(x+a)}$
$= \dfrac{(y+b)\cancel{(x+a)}}{(y+5)\cancel{(x+a)}}$
$= \dfrac{y+b}{y+5}$

11. $\dfrac{x^2+8x+16}{x^2+x-12} \div \dfrac{x^2-16}{x^2-x-6}$
$= \dfrac{x^2+8x+16}{x^2+x-12} \cdot \dfrac{x^2-x-6}{x^2-16}$
$= \dfrac{(x+4)(x+4)}{(x+4)(x-3)} \cdot \dfrac{(x-3)(x+2)}{(x+4)(x-4)}$
$= \dfrac{(x+4)(x+4)(x-3)(x+2)}{(x+4)(x-3)(x+4)(x-4)}$
$= \dfrac{x+2}{x-4}$

13. $\dfrac{3x^2-2x-1}{x^2+6x+8} \div \dfrac{3x^2+13x+4}{x^2+8x+16}$
$= \dfrac{3x^2-2x-1}{x^2+6x+8} \cdot \dfrac{x^2+8x+16}{3x^2+13x+4}$
$= \dfrac{(3x+1)(x-1)}{(x+2)(x+4)} \cdot \dfrac{(x+4)(x+4)}{(3x+1)(x+4)}$
$= \dfrac{(3x+1)(x-1)(x+4)(x+4)}{(x+2)(x+4)(3x+1)(x+4)}$
$= \dfrac{x-1}{x+2}$

15. $\dfrac{x^2}{x-9} - \dfrac{18x-81}{x-9} = \dfrac{x^2-18x+81}{x-9}$
$= \dfrac{(x-9)(x-9)}{x-9}$
$= x-9$

Chapter 7 Review

17. $\dfrac{x}{x+9} + \dfrac{5}{x} = \dfrac{x \cdot x}{(x+9)x} + \dfrac{5(x+9)}{x(x+9)}$

$= \dfrac{x^2 + 5x + 45}{x(x+9)}$

19. $\dfrac{3}{x^2 - 36} - \dfrac{2}{x^2 - 4x - 12}$

$= \dfrac{3}{(x+6)(x-6)} - \dfrac{2}{(x+2)(x-6)}$

$= \dfrac{3(x+2)}{(x+6)(x-6)(x+2)} - \dfrac{2(x+6)}{(x+2)(x-6)(x+6)}$

$= \dfrac{3x + 6 - 2x - 12}{(x+6)(x-6)(x+2)}$

$= \dfrac{x - 6}{(x+6)(x-6)(x+2)}$

$= \dfrac{1}{(x+6)(x+2)}$

21. $\dfrac{3}{x} + \dfrac{1}{2} = \dfrac{5}{x}$

$2x\left(\dfrac{3}{x}\right) + 2x\left(\dfrac{1}{2}\right) = 2x\left(\dfrac{5}{x}\right)$

$6 + x = 10$

$x = 4$

23. $1 - \dfrac{7}{x} = \dfrac{-6}{x^2}$

$x^2(1) + x^2\left(-\dfrac{7}{x}\right) = x^2\left(\dfrac{-6}{x^2}\right)$

$x^2 - 7x = -6$

$x^2 - 7x + 6 = 0$

$(x-1)(x-6) = 0$

$x = 1, 6$

The solutions are 1 and 6.

25. $\dfrac{2}{y^2 - 16} = \dfrac{10}{y^2 + 4y}$

$\dfrac{2}{(y+4)(y-4)} = \dfrac{10}{y(y+4)}$

$y(y+4)(y-4) \cdot \dfrac{2}{(y+4)(y-4)} = y(y+4)(y-4) \cdot \dfrac{10}{y(y+4)}$

$2y = 10y - 40$

$40 = 8y$

$5 = y$

© 2000 Harcourt, Inc

27. Let x = speed of the boat in still water

	d	r	t
upstream	48	$x-3$	$\dfrac{48}{x-3}$
downstream	72	$x+3$	$\dfrac{72}{x+3}$

$$\frac{48}{x-3} = \frac{72}{x+3}$$
$$(x+3)(x-3) \cdot \frac{48}{x-3} = (x+3)(x-3) \cdot \frac{72}{x+3}$$
$$(x+3)48 = (x-3)72$$
$$48x + 144 = 72x - 216$$
$$360 = 24x$$
$$15 = x$$

The speed of the boat in still water is 15 mph.

29. $\dfrac{\frac{x+4}{x^2-16}}{\frac{2}{x-4}} = \dfrac{x+4}{x^2-16} \cdot \dfrac{(x+4)(x-4)}{1}$
$\phantom{29.\ \dfrac{\frac{x+4}{x^2-16}}{\frac{2}{x-4}}} \overline{\phantom{= \dfrac{2}{x-4} \cdot \dfrac{(x+4)(x-4)}{1}}}$
$\phantom{29.\ \dfrac{\frac{x+4}{x^2-16}}{\frac{2}{x-4}}} = \dfrac{2}{x-4} \cdot \dfrac{(x+4)(x-4)}{1}$

$= \dfrac{x+4}{2(x+4)}$

$= \dfrac{1}{2}$

31. $\dfrac{\frac{1}{a-2}+4}{\frac{1}{a-2}+1} = \dfrac{(a-2)\left(\frac{1}{a-2}+4\right)}{(a-2)\left(\frac{1}{a-2}+1\right)}$

$= \dfrac{(a-2)\left(\frac{1}{a-2}\right)+(a-2)(4)}{(a-2)\left(\frac{1}{a-2}\right)+(a-2)(1)}$

$= \dfrac{1+4a-8}{1+a-2}$

$= \dfrac{4a-7}{a-1}$

33. $\dfrac{40 \text{ sec}}{3 \text{ min}} = \dfrac{40}{3(60)} = \dfrac{40}{180} = \dfrac{2}{9}$

35. $\dfrac{a}{3} = \dfrac{12}{a}$
$a^2 = 36$
$a^2 - 36 = 0$
$(a+6)(a-6) = 0$
$a = -6, 6$

The solutions are −6 and 6.

37. $y = kx$: $y = -20$, $x = 4$
$-20 = k(4) \Rightarrow k = -5$
$y = -5x$
If $x = 7$
$y = -5(7) = -35$

Cumulative Review: Chapters 1-7

1. $8 - 11 = 8 + (-11) = -3$

3. $\dfrac{-48}{12} = -4$

5. $5x - 4 - 9x = 5x - 9x - 4 = -4x - 4$

7. $9^{-2} = \dfrac{1}{9^2} = \dfrac{1}{81}$

9. $4^1 + 9^0 + (-7)^0 = 4 + 1 + 1 = 6$

11. $(4a^3 - 10a^2 + 6) - (6a^3 + 5a - 7) = 4a^3 - 10a^2 + 6 - 6a^3 - 5a + 7 = -2a^3 - 10a^2 - 5a + 13$

13. $\dfrac{x^2}{x-7} - \dfrac{14x-49}{x-7} = \dfrac{x^2 - 14x + 49}{x-7} = \dfrac{(x-7)^2}{x-7} = x - 7$

15. $\dfrac{\dfrac{x-2}{x^2+6x+8}}{\dfrac{4}{x+4}} = \dfrac{\dfrac{x-2}{(x+4)(x+2)} \cdot (x+4)(x+2)}{\dfrac{4}{x+4} \cdot (x+4)(x+2)} = \dfrac{x-2}{4(x+2)}$

17. $x - \dfrac{3}{4} = \dfrac{5}{6}$

    $x = \dfrac{3}{4} + \dfrac{5}{6}$

    $x = \dfrac{9}{12} + \dfrac{10}{12}$

    $x = \dfrac{19}{12}$

19. $98r^2 - 18 = 0$

    $2(49r^2 - 9) = 0$

    $2(7r+3)(7r-3) = 0$

    $r = -\dfrac{3}{7}, \dfrac{3}{7}$

21. Multiple each side of the equation by $3x$:

    $3x\left(\dfrac{5}{x} - \dfrac{1}{3}\right) = 3x\left(\dfrac{3}{x}\right)$

    $15 - x = 9$

    $-x = -6$

    $x = 6$

    The solution is 6.

23. $3(x-3) \cdot \dfrac{x}{3} = 3(x-3) \cdot \dfrac{6}{x-3}$

    $x(x-3) = 18$

    $x^2 - 3x = 18$

    $x^2 - 3x - 18 = 0$

    $(x-6)(x+3) = 0$

    $x = 6, -3$

    The solutions are -3 and 6.

25. Multiplying the first equation by -5 and the second equation by 3:
$$-45x - 70y = 20$$
$$45x - 24y = 27$$
Adding the two equations:
$$-94y = 47$$
$$y = -\frac{1}{2}$$
Substituting into the first equation
$$9x + 14\left(-\frac{1}{2}\right) = -4$$
$$9x - 7 = -4$$
$$9x = 3$$
$$x = \frac{1}{3}$$
The solution is $\left(\frac{1}{3}, -\frac{1}{2}\right)$.

27. To clear each equation of fractions, multiply the first equation by 6 and the second equation by 12:
$$6\left(\frac{1}{2}x + \frac{1}{3}y\right) = 6(-1) \qquad 12\left(\frac{1}{3}x\right) = 12\left(\frac{1}{4}y + 5\right)$$
$$3x + 2y = -6 \qquad\qquad 4x = 3y + 60$$
$$\qquad\qquad\qquad\qquad\qquad 4x - 3y = 60$$

$$3x + 2y = -6$$
$$4x - 3y = 60$$
Multiply the first equation by 3 and the second equation by 2:
$$9x + 6y = -18$$
$$8x - 6y = 120$$
Add the two equations:
$$17x = 102$$
$$x = 6$$
Substitute into $3x + 2y = -6$:
$$3(6) + 2y = -6$$
$$18 + 2y = -6$$
$$2y = -24$$
$$y = -12$$
The solution is $(6, -12)$.

29. See the graph in the back of the textbook.

31. $xy + 5x + ay + 5a = x(y + 5) + a(y + 5) = (y + 5)(x + a)$

33. $20y^2 - 27y + 9 = (5y - 3)(4y - 3)$

35. $16x^2 + 72xy + 81y^2 = (4x + 9y)(4x + 9y) = (4x + 9y)^2$

37. Check each ordered pair:
$(3, -1)$: $2(3) - 5 = 6 - 5 = 1 \neq -1$
$(1, -3)$: $2(1) - 5 = 2 - 5 = -3$
$(-2, 9)$: $2(-2) - 5 = -4 - 5 = -9 \neq 9$
The ordered pair $(1, -3)$ is a solution to the equation.

39. $3x - y = 4$ The slope is 3 and the y-intercept is -4.
$-y = -3x + 4$
$y = 3x - 4$

Chapter 7 Test

41. Using the point-slope formula:
$$y-(-1)=-\frac{2}{5}(x-(-2))$$
$$y+1=-\frac{2}{5}(x+2)$$
$$y+1=-\frac{2}{5}x-\frac{4}{5}$$
$$y=-\frac{2}{5}x-\frac{9}{5}$$

43. Associative property of addition

45. $\dfrac{6x-12}{6x+12} \cdot \dfrac{3x+3}{12x-24} = \dfrac{6(x-2)}{6(x+2)} \cdot \dfrac{3(x+1)}{12(x-2)} = \dfrac{18(x-2)(x+1)}{72(x+2)(x-2)} = \dfrac{x+1}{4(x+2)}$

47. $\dfrac{2xy+10x+3y+15}{3xy+15x+2y+10} = \dfrac{2x(y+5)+3(y+5)}{3x(y+5)+2(y+5)} = \dfrac{(y+5)(2x+3)}{(y+5)(3x+2)} = \dfrac{2x+3}{3x+2}$

49. Let x and y = the two numbers.
   $x+y=40$
   $x-y=18$
   Add the two equations:
   $2x=58$
   $x=29$
   Substitute into the first equation:
   $29+y=40$
   $y=11$
   The two numbers are 29 and 11.

Chapter 7 Test

1. $\dfrac{x^2-16}{x^2-8x+16} = \dfrac{(x+4)(x-4)}{(x-4)(x-4)} = \dfrac{x+4}{x-4}$

2. $\dfrac{10a+20}{5a^2+20a+20} = \dfrac{10(a+2)}{5(a+2)(a+2)} = \dfrac{2}{a+2}$

3. $\dfrac{xy+7x+5y+35}{x^2+ax+5x+5a} = \dfrac{x(y+7)+5(y+7)}{x(x+a)+5(x+a)}$
$= \dfrac{(x+5)(y+7)}{(x+5)(x+a)}$
$= \dfrac{y+7}{x+a}$

4. $\dfrac{3x-12}{4} \cdot \dfrac{8}{2x-8} = \dfrac{3(x-4)}{4} \cdot \dfrac{8}{2(x-4)}$
$= \dfrac{24(x-4)}{8(x-4)}$
$= 3$

5. $\dfrac{x^2-49}{x+1} \div \dfrac{x+7}{x^2-1} = \dfrac{x^2-49}{x+1} \cdot \dfrac{x^2-1}{x+7}$

$= \dfrac{(x+7)(x-7)}{x+1} \cdot \dfrac{(x+1)(x-1)}{x+7}$

$= \dfrac{(x+7)(x-7)(x+1)(x-1)}{(x+1)(x+7)}$

$= (x-7)(x-1)$

6. $\dfrac{x^2-3x-10}{x^2-8x+15} \div \dfrac{3x^2+2x-8}{x^2+x-12} = \dfrac{x^2-3x-10}{x^2-8x+15} \cdot \dfrac{x^2+x-12}{3x^2+2x-8}$

$= \dfrac{(x-5)(x+2)}{(x-3)(x-5)} \cdot \dfrac{(x+4)(x-3)}{(x+2)(3x-4)}$

$= \dfrac{x+4}{3x-4}$

7. $(x^2-9)\left(\dfrac{x+2}{x+3}\right) = \dfrac{(x+3)(x-3)}{1}\left(\dfrac{x+2}{x+3}\right) = (x-3)(x+2)$

8. $\dfrac{3}{x-2} - \dfrac{6}{x-2} = \dfrac{3-6}{x-2} = \dfrac{-3}{x-2}$

9. $\dfrac{x}{x^2-9} + \dfrac{4}{4x-12} = \dfrac{x}{(x+3)(x-3)} + \dfrac{4}{4(x-3)}$

$= \dfrac{x}{(x+3)(x-3)} + \dfrac{1}{x-3}$

$= \dfrac{x}{(x+3)(x-3)} + \dfrac{1(x+3)}{(x-3)(x+3)}$

$= \dfrac{2x+3}{(x+3)(x-3)}$

10. $\dfrac{2x}{x^2-1} + \dfrac{x}{x^2-3x+2}$

$= \dfrac{2x}{(x+1)(x-1)} + \dfrac{x}{(x-1)(x-2)}$

$\dfrac{2x(x-2)}{(x+1)(x-1)(x-2)} + \dfrac{x(x+1)}{(x-1)(x-2)(x+1)}$

$= \dfrac{2x^2-4x+x^2+x}{(x+1)(x-1)(x-2)}$

$= \dfrac{3x^2-3x}{(x+1)(x-1)(x-2)}$

$= \dfrac{3x(x-1)}{(x+1)(x-1)(x-2)}$

$= \dfrac{3x}{(x+1)(x-2)}$

11. $\dfrac{7}{5} = \dfrac{x+2}{3}$

$15\left(\dfrac{7}{5}\right) = \left(\dfrac{x+2}{3}\right)15$

$21 = (x+2)5$

$21 = 5x+10$

$11 = 5x$

$\dfrac{11}{5} = x$

Chapter 7 Test

12.
$$\frac{10}{x+4} = \frac{6}{x} - \frac{4}{x}$$
$$x(x+4)\left(\frac{10}{x+4}\right) = x(x+4)\left(\frac{6}{x}\right) - x(x+4)\left(\frac{4}{x}\right)$$
$$10x = 6(x+4) - 4(x+4)$$
$$10x = 6x + 24 - 4x - 16$$
$$10x = 2x + 8$$
$$8x = 8$$
$$x = 1$$

13.
$$\frac{3}{x-2} - \frac{4}{x+1} = \frac{5}{x^2 - x - 2}$$
$$(x+1)(x-2)\left(\frac{3}{x-2}\right) - (x+1)(x-2)\left(\frac{4}{x+1}\right) = (x+1)(x-2)\left(\frac{5}{(x+1)(x-2)}\right)$$
$$3(x+1) - 4(x-2) = 5$$
$$3x + 3 - 4x + 8 = 5$$
$$-x + 11 = 5$$
$$-x = -6$$
$$x = 6$$

14. Let x = the speed of the boat in still water.

	d	r	t
Upstream	26	$x-2$	$\frac{26}{x-2}$
Downstream	34	$x+2$	$\frac{34}{x+2}$

$$\frac{26}{x-2} = \frac{34}{x+2}$$
$$(x+2)(x-2)\left(\frac{26}{x-2}\right) = (x+2)(x-2)\left(\frac{34}{x+2}\right)$$
$$26(x+2) = 34(x-2)$$
$$26x + 52 = 34x - 68$$
$$52 = 8x - 68$$
$$120 = 8x$$
$$15 = x$$

The speed of the boat in still water is 15 mph.

15. Let t = the time to empty the pool with both pipes open.

$$\frac{1}{12} - \frac{1}{15} = \frac{1}{t}$$

$$60t\left(\frac{1}{12} - \frac{1}{15}\right) = 60t \cdot \frac{1}{t}$$

$$5t - 4t = 60$$

$$t = 60$$

It will take 60 hours to empty the pool with both pipes open.

16. $\dfrac{27}{54} = \dfrac{1}{2}$ $\quad\dfrac{\text{solution of alcohol}}{\text{solution of water}}$

$27 + 54 = 81$ $\quad$ total volume

$\dfrac{27}{81} = \dfrac{1}{3}$ $\quad\dfrac{\text{alcohol}}{\text{total volume}}$

17. $\dfrac{8}{100} = \dfrac{x}{1650}$ $\quad\dfrac{\text{defective parts}}{\text{parts produced}}$

$8(1650) = 100x$

$13{,}200 = 100x$

$132 = x$

18. $\dfrac{1 + \dfrac{1}{x}}{1 - \dfrac{1}{x}} = \dfrac{x\left(1 + \dfrac{1}{x}\right)}{x\left(1 - \dfrac{1}{x}\right)}$

$= \dfrac{x(1) + x\left(\dfrac{1}{x}\right)}{x(1) - x\left(\dfrac{1}{x}\right)}$

$= \dfrac{x+1}{x-1}$

19. $\dfrac{1 - \dfrac{16}{x^2}}{1 - \dfrac{2}{x} - \dfrac{8}{x^2}} = \dfrac{x^2\left(1 - \dfrac{16}{x^2}\right)}{x^2\left(1 - \dfrac{2}{x} - \dfrac{8}{x^2}\right)}$

$= \dfrac{x^2(1) - x^2\left(\dfrac{16}{x^2}\right)}{x^2(1) - x^2\left(\dfrac{2}{x}\right) - x^2\left(\dfrac{8}{x^2}\right)}$

$= \dfrac{x^2 - 16}{x^2 - 2x - 8}$

$= \dfrac{(x+4)(x-4)}{(x-4)(x+2)}$

$= \dfrac{x+4}{x+2}$

20. $y = kx^2 : y = 36, x = 3$

$36 = k(3)^2 \Rightarrow k = 4$

$y = 4x^2$

If $x = 5$

$y = 4(5)^2 = 100$

21. $y = \dfrac{k}{x} : y = 6, x = 3$

$6 = \dfrac{k}{3} \Rightarrow k = 18$

$y = \dfrac{18}{x}$

If $x = 9$

$y = \dfrac{18}{9} = 2$

Chapter 8
Roots and Radicals

8.1 Definitions and Common Roots

1. $\sqrt{9} = 3$

3. $-\sqrt{9} = -3$

5. $\sqrt{-25}$ is not a real number

7. $-\sqrt{144} = -12$

9. $\sqrt{625} = 25$

11. $\sqrt{-49}$ is not a real number

13. $-\sqrt{64} = -8$

15. $-\sqrt{100} = -10$

17. $\sqrt{1225} = 35$

19. $\sqrt[4]{1} = 1$

21. $\sqrt[3]{-8} = -2$

23. $-\sqrt[3]{125} = -5$

25. $\sqrt[3]{-1} = -1$

27. $\sqrt[3]{-27} = -3$

29. $-\sqrt[4]{16} = -2$

31. $\sqrt{x^2} = x$

33. $\sqrt{9x^2} = 3x$

35. $\sqrt{x^2 y^2} = xy$

37. $\sqrt{(a+b)^2} = a+b$

39. $\sqrt{49x^2 y^2} = 7xy$

41. $\sqrt[3]{x^3} = x$

43. $\sqrt[3]{8x^3} = 2x$

45. $\sqrt{x^4} = x^2$

47. $\sqrt{36a^6} = 6a^3$

49. $\sqrt{25a^8 b^4} = 5a^4 b^2$

51. $\sqrt[3]{x^6} = x^2$

53. $\sqrt[3]{27a^{12}} = 3a^4$

55. $\sqrt[4]{x^8} = x^2$

57. $\sqrt{9} + \sqrt{16} = 3 + 4 = 7$

59. $\sqrt{9+16} = \sqrt{25} = 5$

61. $\sqrt{144} + \sqrt{25} = 12 + 5 = 17$

63. $\sqrt{144+25} = \sqrt{169} = 13$

65. $\dfrac{5+\sqrt{49}}{2}$ and $\dfrac{5-\sqrt{49}}{2}$

$= \dfrac{5+7}{2} \qquad\quad = \dfrac{5-7}{2}$

$= \dfrac{12}{2} \qquad\qquad = \dfrac{-2}{2}$

$= 6 \qquad\qquad\quad = -1$

67. $\dfrac{2+\sqrt{16}}{2}$ and $\dfrac{52-\sqrt{16}}{2}$

$= \dfrac{2+4}{2} \qquad\quad = \dfrac{2-4}{2}$

$= \dfrac{6}{2} \qquad\qquad = \dfrac{-2}{2}$

$= 3 \qquad\qquad\quad = -1$

© 2000 Harcourt, Inc

Problem Set 8.1

69. $\sqrt{x^2+6x+9} = \sqrt{(x+3)^2} = x+3$

71. See the table in the back of the textbook.

73. $A = \$65 \quad P = \$50$

$r = \dfrac{\sqrt{A}-\sqrt{P}}{\sqrt{P}}$

$r = \dfrac{\sqrt{65}-\sqrt{50}}{\sqrt{50}}$

$r \approx \dfrac{8.062-7.071}{7.071}$

$r = \dfrac{0.991}{7.071}$

$r \approx 0.14 = 14\%$

75. $A = \$600 \quad P = \$500$

$r = \dfrac{\sqrt{A}-\sqrt{P}}{\sqrt{P}}$

$r = \dfrac{\sqrt{600}-\sqrt{500}}{\sqrt{500}}$

$r \approx \dfrac{24.495-22.361}{22.361}$

$r = \dfrac{2.134}{22.361}$

$r \approx 0.095 = 9.5\%$

77. $c = \sqrt{a^2+b^2}; \quad a=4, \ b=3, \ c=x$

$x = \sqrt{4^2+3^2}$

$x = \sqrt{16+9}$

$x = \sqrt{25}$

$x = 5$

79. $c = \sqrt{a^2+b^2}; \quad a=4, \ b=3, \ c=x$

$x = \sqrt{4^2+3^2}$

$x = \sqrt{16+9}$

$x = \sqrt{25}$

$x = 5$

81. $c = \sqrt{a^2+b^2}; \quad a=24 \text{ feet}, \ b=18 \text{ feet}, \ c=x$

$x = \sqrt{(24)^2+(18)^2}$

$x = \sqrt{576+324}$

$x = \sqrt{900}$

$x = 30 \text{ feet}$

83. $c = \sqrt{a^2+b^2}; \quad a=5, \ b=12, \ c=x$

$x = \sqrt{5^2+12^2}$

$x = \sqrt{25+144}$

$x = \sqrt{169}$

$x = 13$

85.
$c = \sqrt{l^2+w^2}; \quad l=3, \ w=4$

$c = \sqrt{3^2+4^2}$

$= \sqrt{9+16}$

$= \sqrt{25}$

$= 5$

$d = \sqrt{c^2+h^2}; \ c=5, \ h=12$

$d = \sqrt{5^2+12^2}$

$= \sqrt{25+144}$

$= \sqrt{169}$

$= 13$

Diagonal $c = 5$ centimeters

Diagonal $d = 13$ centimeters

87. $\dfrac{x^2-16}{x+4} = \dfrac{(x+4)(x-4)}{x+4} = x-4$

© 2000 Harcourt, Inc

Problem Set 8.2

89. $\dfrac{10a+20}{5a^2-20} = \dfrac{10(a+2)}{5(a+2)(a-2)} = \dfrac{2}{a-2}$

91. $\dfrac{2x^2-5x-3}{x^2-3x} = \dfrac{(2x+1)(x-3)}{x(x-3)} = \dfrac{2x+1}{x}$

93. $\dfrac{xy+3x+2y+6}{xy+3x+ay+3a} = \dfrac{x(y+3)+2(y+3)}{x(y+3)+a(y+3)}$
$= \dfrac{(y+3)(x+2)}{(y+3)(x+a)}$
$= \dfrac{x+2}{x+a}$

8.2 Properties of Radicals

1. $\sqrt{8} = \sqrt{4\cdot 2} = \sqrt{4}\cdot\sqrt{2} = 2\sqrt{2}$

3. $\sqrt{12} = \sqrt{4\cdot 3} = \sqrt{4}\sqrt{3} = 2\sqrt{3}$

5. $\sqrt[3]{24} = \sqrt[3]{8\cdot 3} = \sqrt[3]{8}\cdot\sqrt[3]{3} = 2\sqrt[3]{3}$

7. $\sqrt{50x^2} = \sqrt{25x^2\cdot 2} = \sqrt{25x^2}\sqrt{2} = 5x\sqrt{2}$

9. $\sqrt{45a^2b^2} = \sqrt{9a^2b^2\cdot 5} = \sqrt{9a^2b^2}\cdot\sqrt{5} = 3ab\sqrt{5}$

11. $\sqrt[3]{54x^3} = \sqrt[3]{27x^3\cdot 2} = \sqrt[3]{27x^3}\sqrt[3]{2} = 3x\sqrt[3]{2}$

13. $\sqrt{32x^4} = \sqrt{16x^4\cdot 2} = \sqrt{16x^4}\cdot\sqrt{2} = 4x^2\sqrt{2}$

15. $5\sqrt{80} = 5\sqrt{16\cdot 5} = 5\sqrt{16}\sqrt{5} = 5\cdot 4\sqrt{5} = 20\sqrt{5}$

17. $\dfrac{1}{2}\sqrt{28x^3} = \dfrac{1}{2}\sqrt{4x^2\cdot 7x} = \dfrac{1}{2}\sqrt{4x^2}\cdot\sqrt{7x} = \dfrac{1}{2}\cdot 2x\sqrt{7x} = x\sqrt{7x}$

19. $x\sqrt[3]{8x^4} = x\sqrt[3]{8x^3\cdot x} = x\sqrt[3]{8x^3}\sqrt[3]{x} = x\cdot 2x\sqrt[3]{x} = 2x^2\sqrt[3]{x}$

21. $2a\sqrt[3]{27a^5} = 2a\sqrt[3]{27a^3\cdot a^2} = 2a\sqrt[3]{27a^3}\cdot\sqrt[3]{a^2} = 2a\cdot 3a\sqrt[3]{a^2} = 6a^2\sqrt[3]{a^2}$

23. $\dfrac{4}{3}\sqrt{45a^3} = \dfrac{4}{3}\sqrt{9a^2\cdot 5a} = \dfrac{4}{3}\sqrt{9a^2}\sqrt{5a} = \dfrac{4}{3}\cdot 3a\sqrt{5a} = 4a\sqrt{5a}$

25. $3\sqrt{50xy^2} = 3\sqrt{25y^2\cdot 2x} = 3\sqrt{25y^2}\cdot\sqrt{2x} = 3(5y)\sqrt{2x} = 15y\sqrt{2x}$

27. $7\sqrt{12x^2y} = 7\sqrt{4x^2\cdot 3y} = 7\sqrt{4x^2}\sqrt{3y} = 7\cdot 2x\sqrt{3y} = 14x\sqrt{3y}$

29. $\sqrt{\dfrac{16}{25}} = \dfrac{\sqrt{16}}{\sqrt{25}} = \dfrac{4}{5}$

31. $\sqrt{\dfrac{4}{9}} = \dfrac{\sqrt{4}}{\sqrt{9}} = \dfrac{2}{3}$

33. $\sqrt[3]{\dfrac{8}{27}} = \dfrac{\sqrt[3]{8}}{\sqrt[3]{27}} = \dfrac{2}{3}$

35. $\sqrt[4]{\dfrac{16}{81}} = \dfrac{\sqrt[4]{16}}{\sqrt[4]{81}} = \dfrac{2}{3}$

37. $\sqrt{\dfrac{100x^2}{25}} = \dfrac{\sqrt{100x^2}}{\sqrt{25}} = \dfrac{10x}{5} = 2x$

39. $\sqrt{\dfrac{81a^2b^2}{9}} = \sqrt{9a^2b^2} = 3ab$

41. $\sqrt[3]{\dfrac{27x^3}{8y^3}} = \dfrac{\sqrt[3]{27x^3}}{\sqrt[3]{8y^3}} = \dfrac{3x}{2y}$

43. $\sqrt{\dfrac{50}{9}} = \dfrac{\sqrt{50}}{\sqrt{9}} = \dfrac{\sqrt{25 \cdot 2}}{3} = \dfrac{5\sqrt{2}}{3}$

45. $\sqrt{\dfrac{75}{25}} = \sqrt{3}$

47. $\sqrt{\dfrac{128}{49}} = \dfrac{\sqrt{128}}{\sqrt{49}} = \dfrac{\sqrt{64 \cdot 2}}{7} = \dfrac{8\sqrt{2}}{7}$

49. $\sqrt{\dfrac{288x}{25}} = \dfrac{\sqrt{288x}}{\sqrt{25}} = \dfrac{\sqrt{144 \cdot 2x}}{\sqrt{25}} = \dfrac{\sqrt{144} \cdot \sqrt{2x}}{\sqrt{25}} = \dfrac{12\sqrt{2x}}{5}$

51. $\sqrt{\dfrac{54a^2}{25}} = \dfrac{\sqrt{54a^2}}{\sqrt{25}} = \dfrac{\sqrt{9a^2 \cdot 6}}{5} = \dfrac{3a\sqrt{6}}{5}$

53. $\dfrac{3\sqrt{50}}{2} = \dfrac{3\sqrt{25}\sqrt{2}}{2} = \dfrac{3(5)\sqrt{2}}{2} = \dfrac{15\sqrt{2}}{2}$

55. $\dfrac{7\sqrt{28y^2}}{3} = \dfrac{7\sqrt{4y^2 \cdot 7}}{3} = \dfrac{7 \cdot 2y\sqrt{7}}{3} = \dfrac{14y\sqrt{7}}{3}$

57. $\dfrac{5\sqrt{72a^2b^2}}{\sqrt{36}} = \dfrac{5\sqrt{36a^2b^2 \cdot 2}}{\sqrt{36}} = \dfrac{5\sqrt{36a^2b^2} \cdot \sqrt{2}}{\sqrt{36}} = \dfrac{5(6)ab\sqrt{2}}{6} = 5ab\sqrt{2}$

59. $\dfrac{6\sqrt{8x^2y}}{\sqrt{4}} = \dfrac{6\sqrt{4x^2 \cdot 2y}}{2} = \dfrac{6 \cdot 2x\sqrt{2y}}{2} = \dfrac{12x\sqrt{2y}}{2} = 6x\sqrt{2y}$

61. See the table in the back of the textbook.

63. See the table in the back of the textbook.

65. Let $h = 25$ feet, then

$t = \sqrt{\dfrac{h}{16}}$

$t = \sqrt{\dfrac{25}{16}}$

$t = \dfrac{5}{4}$ seconds

67. $\dfrac{8x}{x^2 - 5x} \cdot \dfrac{x^2 - 25}{4x^2 + 4x} = \dfrac{8x}{x(x-5)} \cdot \dfrac{(x+5)(x-5)}{4x(x+1)}$

$= \dfrac{8x(x+5)(x-5)}{4x^2(x-5)(x+1)} = \dfrac{2(x+5)}{x(x+1)}$

69. $\dfrac{x^2 + 3x - 4}{3x^2 + 7x - 20} \div \dfrac{x^2 - 2x + 1}{3x^2 - 2x - 5} = \dfrac{x^2 + 3x - 4}{3x^2 + 7x - 20} \cdot \dfrac{3x^2 - 2x - 5}{x^2 - 2x + 1}$

$= \dfrac{(x+4)(x-1)(x+1)(3x-5)}{(x+4)(3x-5)(x-1)(x-1)}$

$= \dfrac{x+1}{x-1}$

Problem Set 8.3

71. $(x^2-36)\left(\dfrac{x+3}{x-6}\right) = \dfrac{(x+6)(x-6)}{1} \cdot \dfrac{x+3}{x-6} = \dfrac{(x+6)(x-6)(x+3)}{x-6} = (x+6)(x+3)$

8.3 Simplified Form for Radicals

1. $\sqrt{\dfrac{1}{2}} = \dfrac{\sqrt{1}}{\sqrt{2}} \cdot \dfrac{\sqrt{2}}{\sqrt{2}} = \dfrac{\sqrt{2}}{\sqrt{4}} = \dfrac{\sqrt{2}}{2}$

3. $\sqrt{\dfrac{1}{3}} = \dfrac{\sqrt{1}}{\sqrt{3}} \cdot \dfrac{\sqrt{3}}{\sqrt{3}} = \dfrac{\sqrt{3}}{\sqrt{9}} = \dfrac{\sqrt{3}}{3}$

5. $\sqrt{\dfrac{2}{5}} = \dfrac{\sqrt{2}}{\sqrt{5}} \cdot \dfrac{\sqrt{5}}{\sqrt{5}} = \dfrac{\sqrt{10}}{\sqrt{25}} = \dfrac{\sqrt{10}}{5}$

7. $\sqrt{\dfrac{3}{2}} = \dfrac{\sqrt{3}}{\sqrt{2}} \cdot \dfrac{\sqrt{2}}{\sqrt{2}} = \dfrac{\sqrt{6}}{\sqrt{4}} = \dfrac{\sqrt{6}}{2}$

9. $\sqrt{\dfrac{20}{3}} = \dfrac{\sqrt{20}}{\sqrt{3}} \cdot \dfrac{\sqrt{3}}{\sqrt{3}} = \dfrac{\sqrt{60}}{\sqrt{9}} = \dfrac{\sqrt{4\cdot 15}}{3} = \dfrac{2\sqrt{15}}{3}$

11. $\sqrt{\dfrac{45}{6}} = \sqrt{\dfrac{15}{2}} = \dfrac{\sqrt{15}}{\sqrt{2}} \cdot \dfrac{\sqrt{2}}{\sqrt{2}} = \dfrac{\sqrt{30}}{\sqrt{4}} = \dfrac{\sqrt{30}}{2}$

13. $\sqrt{\dfrac{20}{5}} = \sqrt{4} = 2$

15. $\dfrac{\sqrt{21}}{\sqrt{3}} = \sqrt{\dfrac{21}{3}} = \sqrt{7}$

17. $\dfrac{\sqrt{35}}{\sqrt{7}} = \sqrt{\dfrac{35}{7}} = \sqrt{5}$

19. $\dfrac{10\sqrt{15}}{5\sqrt{3}} = \dfrac{10}{5} \cdot \sqrt{\dfrac{15}{3}} = 2\sqrt{5}$

21. $\dfrac{6\sqrt{21}}{3\sqrt{7}} = \dfrac{6}{3} \cdot \sqrt{\dfrac{21}{7}} = 2\sqrt{3}$

23. $\dfrac{6\sqrt{35}}{12\sqrt{5}} = \dfrac{6}{12} \cdot \sqrt{\dfrac{35}{5}} = \dfrac{1}{2}\sqrt{7} = \dfrac{\sqrt{7}}{2}$

25. $\sqrt{\dfrac{4x^2y^2}{2}} = \sqrt{2x^2y^2} = xy\sqrt{2}$

27. $\sqrt{\dfrac{5x^2y}{3}} = \dfrac{\sqrt{5x^2y}}{\sqrt{3}} \cdot \dfrac{\sqrt{3}}{\sqrt{3}} = \dfrac{\sqrt{15x^2y}}{\sqrt{9}} = \dfrac{x\sqrt{15y}}{3}$

29. $\sqrt{\dfrac{16a^4}{5}} = \dfrac{\sqrt{16a^4}}{\sqrt{5}} = \dfrac{4a^2}{\sqrt{5}} \cdot \dfrac{\sqrt{5}}{\sqrt{5}} = \dfrac{4a^2\sqrt{5}}{5}$

31. $\sqrt{\dfrac{72a^5}{5}} = \dfrac{\sqrt{72a^5}}{\sqrt{5}} \cdot \dfrac{\sqrt{5}}{\sqrt{5}} = \dfrac{\sqrt{360a^5}}{\sqrt{25}} = \dfrac{\sqrt{36a^4 \cdot 10a}}{5} = \dfrac{6a^2\sqrt{10a}}{5}$

33. $\sqrt{\dfrac{20x^2y^3}{3}} = \dfrac{\sqrt{4x^2y^2 \cdot 5y}}{\sqrt{3}} = \dfrac{2xy\sqrt{5y}}{\sqrt{3}} \cdot \dfrac{\sqrt{3}}{\sqrt{3}} = \dfrac{2xy\sqrt{15y}}{3}$

35. $\dfrac{2\sqrt{20x^2y^3}}{3} = \dfrac{2\sqrt{4x^2y^2 \cdot 5y}}{3} = \dfrac{2 \cdot 2xy\sqrt{5y}}{3} = \dfrac{4xy\sqrt{5y}}{3}$

37. $\dfrac{6\sqrt{54a^2b^3}}{5} = \dfrac{6\sqrt{9a^2b^2 \cdot 6b}}{5} = \dfrac{6(3)ab\sqrt{6b}}{5} = \dfrac{18ab\sqrt{6b}}{5}$

© 2000 Harcourt, Inc

39. $\dfrac{3\sqrt{72x^4}}{\sqrt{2x}} = 3\sqrt{\dfrac{72x^4}{2x}} = 3\sqrt{36x^3} = 3\sqrt{36x^2 \cdot x} = 3 \cdot 6x\sqrt{x} = 18x\sqrt{x}$

41. $\sqrt[3]{\dfrac{1}{2}} = \dfrac{\sqrt[3]{1}}{\sqrt[3]{2}} \cdot \dfrac{\sqrt[3]{4}}{\sqrt[3]{4}} = \dfrac{\sqrt[3]{4}}{\sqrt[3]{8}} = \dfrac{\sqrt[3]{4}}{2}$

43. $\sqrt[3]{\dfrac{1}{9}} = \dfrac{\sqrt[3]{1}}{\sqrt[3]{9}} \cdot \dfrac{\sqrt[3]{3}}{\sqrt[3]{3}} = \dfrac{\sqrt[3]{3}}{\sqrt[3]{27}} = \dfrac{\sqrt[3]{3}}{3}$

45. $\sqrt[3]{\dfrac{3}{2}} = \dfrac{\sqrt[3]{3}}{\sqrt[3]{2}} \cdot \dfrac{\sqrt[3]{4}}{\sqrt[3]{4}}$
$= \dfrac{\sqrt[3]{12}}{\sqrt[3]{8}}$
$= \dfrac{\sqrt[3]{12}}{2}$

47. See the table in the back of the textbook.

49. See the table in the back of the textbook.

51. Substituting $h = 24$: $d = \sqrt{\dfrac{3 \cdot 24}{2}} = \sqrt{36} = 6$ miles

53. See the drawing in the back of the textbook.

55. $\sqrt{1^2 + 1} = \sqrt{1+1} = \sqrt{2}$
$\sqrt{(\sqrt{2})^2 + 1} = \sqrt{2+1} = \sqrt{3}$
$\sqrt{(\sqrt{3})^2 + 1} = \sqrt{3+1} = \sqrt{4} = 2$

57. $3x + 7x = (3+7)x = 10x$

59. $15x + 8x = (15+8)x = 23x$

61. $7a - 3a + 6a = (7-3+6)a = 10a$

63. $\dfrac{x^2}{x+5} + \dfrac{10x+25}{x+5} = \dfrac{x^2 + 10x + 25}{x+5} = \dfrac{(x+5)^2}{x+5} = x+5$

65. $\dfrac{a}{3} + \dfrac{2}{5} = \dfrac{a}{3} \cdot \dfrac{5}{5} + \dfrac{2}{5} \cdot \dfrac{3}{3} = \dfrac{5a}{15} + \dfrac{6}{15} = \dfrac{5a+6}{15}$

67. $\dfrac{6}{a^2-9} - \dfrac{5}{a^2-a-6} = \dfrac{6}{(a+3)(a-3)} - \dfrac{5}{(a-3)(a+2)}$
$= \dfrac{6 \cdot (a+2)}{(a+3)(a-3)(a+2)} - \dfrac{5 \cdot (a+3)}{(a+3)(a-3)(a+2)}$
$= \dfrac{6a+12}{(a+3)(a-3)(a+2)} - \dfrac{5a+15}{(a+3)(a-3)(a+2)}$
$= \dfrac{6a+12-5a-15}{(a+3)(a-3)(a+2)}$
$= \dfrac{a-3}{(a+3)(a-3)(a+2)}$
$= \dfrac{1}{(a+3)(a+2)}$

8.4 Addition and Subtraction of Radical Expressions

1. $3\sqrt{2} + 4\sqrt{2} = 7\sqrt{2}$

3. $9\sqrt{5} - 7\sqrt{5} = 2\sqrt{5}$

5. $\sqrt{3} + 6\sqrt{3} = 7\sqrt{3}$

7. $\frac{5}{8}\sqrt{5} - \frac{3}{7}\sqrt{5} = \frac{35}{56}\sqrt{5} - \frac{24}{56}\sqrt{5} = \frac{11}{56}\sqrt{5}$

9. $14\sqrt{13} - \sqrt{13} = 13\sqrt{13}$

11. $-3\sqrt{10} + 9\sqrt{10} = 6\sqrt{10}$

13. $5\sqrt{5} + \sqrt{5} = 6\sqrt{5}$

15. $\sqrt{8} + 2\sqrt{2} = \sqrt{4 \cdot 2} + 2\sqrt{2} = 2\sqrt{2} + 2\sqrt{2} = 4\sqrt{2}$

17. $3\sqrt{3} - \sqrt{27} = 3\sqrt{3} - \sqrt{9 \cdot 3} = 3\sqrt{3} - 3\sqrt{3} = 0$

19. $5\sqrt{12} - 10\sqrt{48} = 5\sqrt{4 \cdot 3} - 10\sqrt{16 \cdot 3} = 5 \cdot 2\sqrt{3} - 10 \cdot 4\sqrt{3} = 10\sqrt{3} - 40\sqrt{3} = -30\sqrt{3}$

21. $-\sqrt{75} - \sqrt{3} = -\sqrt{25 \cdot 3} - \sqrt{3} = -5\sqrt{3} - \sqrt{3} = -6\sqrt{3}$

23. $\frac{1}{5}\sqrt{75} - \frac{1}{2}\sqrt{12} = \frac{1}{5}\sqrt{25 \cdot 3} - \frac{1}{2}\sqrt{4 \cdot 3} = \frac{1}{5} \cdot 5\sqrt{3} - \frac{1}{2} \cdot 2\sqrt{3} = \sqrt{3} - \sqrt{3} = 0$

25. $\frac{3}{4}\sqrt{8} + \frac{3}{10}\sqrt{75} = \frac{3}{4}\sqrt{4 \cdot 2} + \frac{3}{10}\sqrt{25 \cdot 3} = \frac{3}{4} \cdot 2\sqrt{2} + \frac{3}{10} \cdot 5\sqrt{3} = \frac{3}{2}\sqrt{2} + \frac{3}{2}\sqrt{3}$

27. $\sqrt{27} - 2\sqrt{12} + \sqrt{3} = \sqrt{9 \cdot 3} - 2\sqrt{4 \cdot 3} + \sqrt{3} = 3\sqrt{3} - 4\sqrt{3} + \sqrt{3} = 0$

29. $\frac{5}{6}\sqrt{72} - \frac{3}{8}\sqrt{8} + \frac{3}{10}\sqrt{50} = \frac{5}{6}\sqrt{36 \cdot 2} - \frac{3}{8}\sqrt{4 \cdot 2} + \frac{3}{10}\sqrt{25 \cdot 2}$

$= \frac{5}{6} \cdot 6\sqrt{2} - \frac{3}{8} \cdot 2\sqrt{2} + \frac{3}{10} \cdot 5\sqrt{2}$

$= 5\sqrt{2} - \frac{3}{4}\sqrt{2} + \frac{3}{2}\sqrt{2}$

$= \left(\frac{20}{4} - \frac{3}{4} + \frac{6}{4}\right)\sqrt{2}$

$= \frac{23}{4}\sqrt{2}$

31. $5\sqrt{7} + 2\sqrt{28} - 4\sqrt{63} = 5\sqrt{7} + 2\sqrt{4 \cdot 7} - 4\sqrt{9 \cdot 7} = 5\sqrt{7} + 2 \cdot 2\sqrt{7} - 4 \cdot 3\sqrt{7} = 5\sqrt{7} + 4\sqrt{7} - 12\sqrt{7} = -3\sqrt{7}$

33. $6\sqrt{48} - 2\sqrt{12} + 5\sqrt{27} = 6\sqrt{16 \cdot 3} - 2\sqrt{4 \cdot 3} + 5\sqrt{9 \cdot 3}$

$= 6 \cdot 4\sqrt{3} - 2 \cdot 2\sqrt{3} + 5 \cdot 3\sqrt{3}$

$= 24\sqrt{3} - 4\sqrt{3} + 15\sqrt{3}$

$= 35\sqrt{3}$

35. $6\sqrt{48} - \sqrt{72} - 3\sqrt{300} = 6\sqrt{16 \cdot 3} - \sqrt{36 \cdot 2} - 3\sqrt{100 \cdot 3}$
$= 6 \cdot 4\sqrt{3} - 6\sqrt{2} - 3 \cdot 10\sqrt{3}$
$= 24\sqrt{3} - 6\sqrt{2} - 30\sqrt{3}$
$= -6\sqrt{3} - 6\sqrt{2}$

37. $\sqrt{x^3} + x\sqrt{x} = \sqrt{x^2 \cdot x} + x\sqrt{x}$
$= x\sqrt{x} + x\sqrt{x}$
$= 2x\sqrt{x}$

39. $5\sqrt{3a^2} - a\sqrt{3} = 5a\sqrt{3} - a\sqrt{3} = 4a\sqrt{3}$

41. $5\sqrt{8x^3} + x\sqrt{50x} = 5\sqrt{4x^2 \cdot 2x} + x\sqrt{25 \cdot 2x} = 5 \cdot 2x\sqrt{2x} + x \cdot 5\sqrt{2x} = 10x\sqrt{2x} + 5x\sqrt{2x} = 15x\sqrt{2x}$

43. $3\sqrt{75x^3 y} - 2x\sqrt{3xy} = 3\sqrt{25x^2 \cdot 3xy} - 2x\sqrt{3xy} = 3 \cdot 5x\sqrt{3xy} - 2x\sqrt{3xy} = 15x\sqrt{3xy} - 2x\sqrt{3xy} = 13x\sqrt{3xy}$

45. $\sqrt{20ab^2} - b\sqrt{45a} = \sqrt{4b^2 \cdot 5a} - b\sqrt{9 \cdot 5a} = 2b\sqrt{5a} - 3b\sqrt{5a} = -b\sqrt{5a}$

47. $9\sqrt{18x^3} - 2x\sqrt{48x} = 9\sqrt{9x^2 \cdot 2x} - 2x\sqrt{16 \cdot 3x} = 9 \cdot 3x\sqrt{2x} - 2x \cdot 4\sqrt{3x} = 27x\sqrt{2x} - 8x\sqrt{3x}$

49. $7\sqrt{50x^2 y} + 8x\sqrt{8y} - 7\sqrt{32x^2 y} = 7\sqrt{25x^2 \cdot 2y} + 8x\sqrt{4 \cdot 2y} - 7\sqrt{16x^2 \cdot 2y}$
$= 7 \cdot 5x\sqrt{2y} + 8x \cdot 2\sqrt{2y} - 7 \cdot 4x\sqrt{2y}$
$= 35x\sqrt{2y} + 16x\sqrt{2y} - 28x\sqrt{2y}$
$= 23x\sqrt{2y}$

51. $\dfrac{8-\sqrt{24}}{6} = \dfrac{8-\sqrt{4 \cdot 6}}{6} = \dfrac{8-2\sqrt{6}}{6} = \dfrac{2(4-\sqrt{6})}{6} = \dfrac{4-\sqrt{6}}{3}$

53. $\dfrac{6+\sqrt{8}}{2} = \dfrac{6+\sqrt{4 \cdot 2}}{2} = \dfrac{6+2\sqrt{2}}{2} = \dfrac{2(3+\sqrt{2})}{2} = 3+\sqrt{2}$

55. $\dfrac{-10+\sqrt{50}}{10} = \dfrac{-10+\sqrt{25 \cdot 2}}{10} = \dfrac{-10+5\sqrt{2}}{10} = \dfrac{5(-2+\sqrt{2})}{10} = \dfrac{-2+\sqrt{2}}{2}$

57. See the table in the back of the textbook.

59. See the table in the back of the textbook.

61. The correct statement is: $4\sqrt{3} + 5\sqrt{3} = 9\sqrt{3}$

63. $(3x+y)^2 = (3x)^2 + 2(3x)(y) + y^2 = 9x^2 + 6xy + y^2$

65. $(3x-4y)(3x+4y) = (3x)^2 - (4y)^2 = 9x^2 - 16y^2$

67. $6\left(\dfrac{x}{3}-\dfrac{1}{2}\right) = 6\left(\dfrac{5}{2}\right)$
$2x - 3 = 15$
$2x = 18$
$x = 9$
The solution is $x = 9$.

Problem Set 8.5

69. $x^2\left(1-\dfrac{5}{x}\right) = x^2\left(\dfrac{-6}{x^2}\right)$

$x^2 - 5x = -6$

$x^2 - 5x + 6 = 0$

$(x-2)(x-3) = 0$

$x = 2, 3$

The solutions are $x = 2$ and $x = 3$.

71. $2(a-4)\left(\dfrac{a}{a-4} - \dfrac{a}{2}\right) = 2(a-4) \cdot \dfrac{4}{a-4}$

$2a - a(a-4) = 8$

$2a - a^2 + 4a = 8$

$-a^2 + 6a - 8 = 0$

$a^2 - 6a + 8 = 0$

$(a-4)(a-2) = 0$

$a = 2, 4$

Since $a = 4$ does not check in the original equation, the solution is $a = 2$.

8.5 Multiplication and Division of Radicals

1. $\sqrt{3}\sqrt{2} = \sqrt{6}$

3. $\sqrt{6}\sqrt{2} = \sqrt{12} = \sqrt{4 \cdot 3} = 2\sqrt{3}$

5. $(2\sqrt{3})(5\sqrt{7}) = 10\sqrt{21}$

7. $(4\sqrt{3})(2\sqrt{6}) = 8\sqrt{18} = 8\sqrt{9 \cdot 2} = 8 \cdot 3\sqrt{2} = 24\sqrt{2}$

9. $\sqrt{2}(\sqrt{3} - 1) = \sqrt{6} - \sqrt{2}$

11. $\sqrt{2}(\sqrt{3} + \sqrt{2}) = \sqrt{6} + \sqrt{4} = \sqrt{6} + 2$

13. $\sqrt{3}(2\sqrt{2} + \sqrt{3}) = 2\sqrt{6} + \sqrt{9} = 2\sqrt{6} + 3$

15. $\sqrt{3}(2\sqrt{3} - \sqrt{5}) = 2\sqrt{9} - \sqrt{15} = 2 \cdot 3 - \sqrt{15} = 6 - \sqrt{15}$

17. $2\sqrt{3}(\sqrt{2} + \sqrt{5}) = 2\sqrt{6} + 2\sqrt{15}$

19. $(\sqrt{2} + 1)^2 = (\sqrt{2})^2 + 2(\sqrt{2})(1) + (1)^2 = 2 + 2\sqrt{2} + 1 = 3 + 2\sqrt{2}$

21. $(\sqrt{x} + 3)^2 = (\sqrt{x})^2 + 2(\sqrt{x})(3) + (3)^2 = x + 6\sqrt{x} + 9$

23. $(5 - \sqrt{2})^2 = (5)^2 - 2(5)(\sqrt{2}) + (\sqrt{2})^2 = 25 - 10\sqrt{2} + 2 = 27 - 10\sqrt{2}$

25. $\left(\sqrt{a} - \dfrac{1}{2}\right)^2 = (\sqrt{a})^2 - 2(\sqrt{a})\left(\dfrac{1}{2}\right) + \left(\dfrac{1}{2}\right)^2 = a - \sqrt{a} + \dfrac{1}{4}$

27. $(3 + \sqrt{7})^2 = (3)^2 + 2(3)(\sqrt{7}) + (\sqrt{7})^2 = 9 + 6\sqrt{7} + 7 = 16 + 6\sqrt{7}$

29. $(\sqrt{5} + 3)(\sqrt{5} + 2) = 5 + 3\sqrt{5} + 2\sqrt{5} + 6 = 11 + 5\sqrt{5}$

31. $(\sqrt{2} - 5)(\sqrt{2} + 6) = 2 - 5\sqrt{2} + 6\sqrt{2} - 30 = -28 + \sqrt{2}$

33. $\left(\sqrt{3}+\dfrac{1}{2}\right)\left(\sqrt{2}+\dfrac{1}{3}\right) = \sqrt{6}+\dfrac{1}{2}\sqrt{2}+\dfrac{1}{3}\sqrt{3}+\dfrac{1}{6}$

35. $\left(\sqrt{x}+6\right)\left(\sqrt{x}-6\right) = \left(\sqrt{x}\right)^2 - (6)^2 = x-36$

37. $\left(\sqrt{a}+\dfrac{1}{3}\right)\left(\sqrt{a}+\dfrac{2}{3}\right) = a+\dfrac{1}{3}\sqrt{a}+\dfrac{2}{3}\sqrt{a}+\dfrac{2}{9} = a+\sqrt{a}+\dfrac{2}{9}$

39. $\left(\sqrt{5}-2\right)\left(\sqrt{5}+2\right) = \left(\sqrt{5}\right)^2 - (2)^2 = 5-4 = 1$

41. $\left(2\sqrt{7}+3\right)\left(3\sqrt{7}-4\right) = 6\sqrt{49}+9\sqrt{7}-8\sqrt{7}-12 = 42+\sqrt{7}-12 = 30+\sqrt{7}$

43. $\left(2\sqrt{x}+4\right)\left(3\sqrt{x}+2\right) = 6x+12\sqrt{x}+4\sqrt{x}+8 = 6x+16\sqrt{x}+8$

45. $\left(7\sqrt{a}+2\sqrt{b}\right)\left(7\sqrt{a}-2\sqrt{b}\right) = \left(7\sqrt{a}\right)^2 - \left(2\sqrt{b}\right)^2 = 49a-4b$

47. $\dfrac{\sqrt{3}}{\sqrt{5}-\sqrt{2}} \cdot \dfrac{\sqrt{5}+\sqrt{2}}{\sqrt{5}+\sqrt{2}} = \dfrac{\sqrt{15}+\sqrt{6}}{5-2} = \dfrac{\sqrt{15}+\sqrt{6}}{3}$

49. $\dfrac{\sqrt{5}}{\sqrt{5}+\sqrt{2}} \cdot \dfrac{\sqrt{5}-\sqrt{2}}{\sqrt{5}-\sqrt{2}} = \dfrac{\sqrt{25}-\sqrt{10}}{5-2} = \dfrac{5-\sqrt{10}}{3}$

51. $\dfrac{8}{3-\sqrt{5}} \cdot \dfrac{3+\sqrt{5}}{3+\sqrt{5}} = \dfrac{8(3+\sqrt{5})}{9-5} = \dfrac{8(3+\sqrt{5})}{4} = 2(3+\sqrt{5}) = 6+2\sqrt{5}$

53. $\dfrac{\sqrt{3}+\sqrt{2}}{\sqrt{3}-\sqrt{2}} \cdot \dfrac{\sqrt{3}+\sqrt{2}}{\sqrt{3}+\sqrt{2}} = \dfrac{3+\sqrt{6}+\sqrt{6}+2}{3-2} = \dfrac{5+2\sqrt{6}}{1} = 5+2\sqrt{6}$

55. $\dfrac{\sqrt{7}-\sqrt{3}}{\sqrt{7}+\sqrt{3}} \cdot \dfrac{\sqrt{7}-\sqrt{3}}{\sqrt{7}-\sqrt{3}} = \dfrac{7-\sqrt{21}-\sqrt{21}+3}{7-3} = \dfrac{10-2\sqrt{21}}{4} = \dfrac{2(5-\sqrt{21})}{4} = \dfrac{5-\sqrt{21}}{2}$

57. $\dfrac{\sqrt{x}+2}{\sqrt{x}-2} \cdot \dfrac{\sqrt{x}+2}{\sqrt{x}+2} = \dfrac{x+2\sqrt{2}+2\sqrt{x}+4}{x-4} = \dfrac{x+4\sqrt{x}+4}{x-4}$

59. $\dfrac{\sqrt{5}-\sqrt{2}}{\sqrt{5}+\sqrt{3}} \cdot \dfrac{\sqrt{5}-\sqrt{3}}{\sqrt{5}-\sqrt{3}} = \dfrac{5-\sqrt{10}-\sqrt{15}+\sqrt{6}}{5-3} = \dfrac{5-\sqrt{10}-\sqrt{15}+\sqrt{6}}{2}$

61. The correct statement is: $2(3\sqrt{5}) = 6\sqrt{5}$

63. The correct statement is: $(\sqrt{3}+7)^2 = (\sqrt{3})^2 + 2(\sqrt{3})(7) + (7)^2 = 3+14\sqrt{3}+49 = 52+14\sqrt{3}$

Problem Set 8.6

65. $x^2 + 5x - 6 = 0$
$(x+6)(x-1) = 0$
$x = -6, 1$

67. $x^2 - 3x = 0$
$x(x-3) = 0$
$x = 0, 3$

69. $\dfrac{x}{3} = \dfrac{27}{x}$
$x^2 = 81$
$x^2 - 81 = 0$
$(x+9)(x-9) = 0$
$x = -9, 9$

71. $\dfrac{x}{5} = \dfrac{3}{x+2}$
$x^2 + 2x = 15$
$x^2 + 2x - 15 = 0$
$(x+5)(x-3) = 0$
$x = -5, 3$

73. Comparing miles to hours
$\dfrac{375}{15} = \dfrac{x}{20}$
$15x = 7500$
$x = 500$
You will drive 500 miles.

8.6 Equations Involving Radicals

1. $\sqrt{x+1} = 2$
$\left(\sqrt{x+1}\right)^2 = 2^2$
$x + 1 = 4$
$x = 3$
The solution is $x = 3$.

3. $\sqrt{x+5} = 7$
$\left(\sqrt{x+5}\right)^2 = (7)^2$
$x + 5 = 49$
$x = 44$
The solution is $x = 44$.

5. $\sqrt{x-9} = -6$
$\left(\sqrt{x-9}\right)^2 = (-6)^2$
$x - 9 = 36$
$x = 45$
Does not check. There is no solution.

7. $\sqrt{x-5} = -4$
$\left(\sqrt{x-5}\right)^2 = (-4)^2$
$x - 5 = 16$
$x = 21$
Does not check. There is no solution.

9. $\sqrt{x-8} = 0$
$\left(\sqrt{x-8}\right)^2 = 0^2$
$x - 8 = 0$
$x = 8$
The solution is $x = 8$.

11. $\sqrt{2x+1} = 3$
$\left(\sqrt{2x+1}\right)^2 = (3)^2$
$2x + 1 = 9$
$x = 4$
The solution is $x = 4$.

13. $\sqrt{2x-3} = -5$
$\left(\sqrt{2x-3}\right)^2 = (-5)^2$
$2x - 3 = 25$
$2x = 28$
$x = 14$

Does not check. There is no solution.

15. $\sqrt{3x+6} = 2$
$\left(\sqrt{3x+6}\right)^2 = (2)^2$
$3x + 6 = 4$
$x = -\dfrac{2}{3}$

The solution is $x = -\dfrac{2}{3}$.

17. $2\sqrt{x} = 10$
$\sqrt{x} = 5$
$\left(\sqrt{x}\right)^2 = 5^2$
$x = 25$

The solution is $x = 25$.

19. $3\sqrt{a} = 6$
$\sqrt{a} = 2$
$\left(\sqrt{a}\right)^2 = (2)^2$
$a = 4$

The solution is $a = 4$.

21. $\sqrt{3x+4} - 3 = 2$
$\sqrt{3x+4} = 5$
$\left(\sqrt{3x+4}\right)^2 = 5^2$
$3x + 4 = 25$
$3x = 21$
$x = 7$

The solution is $x = 7$.

23. $\sqrt{5y-4} - 2 = 4$
$\sqrt{5y-4} = 6$
$\left(\sqrt{5y-4}\right)^2 = (6)^2$
$5y - 4 = 36$
$5y = 40$
$y = 8$

The solution is $y = 8$.

25. $\sqrt{2x+1} + 5 = 2$
$\sqrt{2x+1} = -3$
$\left(\sqrt{2x+1}\right)^2 = (-3)^2$
$2x + 1 = 9$
$2x = 8$
$x = 4$

Does not check. There is no solution.

27. $\sqrt{x+3} = x - 3$
$\left(\sqrt{x+3}\right)^2 = (x-3)^2$
$x + 3 = x^2 - 6x + 9$
$0 = x^2 - 7x + 6$
$0 = (x-6)(x-1)$
$x = 6, 1$

Possible solutions are 6 and 1, only 6 checks.

29. $\sqrt{a+2} = a + 2$
$\left(\sqrt{a+2}\right)^2 = (a+2)^2$
$a + 2 = a^2 + 4a + 4$
$0 = a^2 + 3a + 2$
$0 = (a+2)(a+1)$

The solutions are -2 and -1.

31. $\sqrt{2x+9} = x + 5$
$\left(\sqrt{2x+9}\right)^2 = (x+5)^2$
$2x + 9 = x^2 + 10x + 25$
$0 = x^2 + 8x + 16$
$0 = (x+4)^2$
$x = -4$

The solution is $x = -4$.

Problem Set 8.6

33. $\sqrt{y-4} = y-6$
$\left(\sqrt{y-4}\right)^2 = (y-6)^2$
$y-4 = y^2 -12y+36$
$0 = y^2 -13y+40$
$0 = (y-8)(y-5)$
$y = 8, 5$

35. See the graph in the back of the textbook.

37. $y = \sqrt{x}$ See the graph in the back of the textbook.

39. $y = 2\sqrt{x}$ See the graph in the back of the textbook.

41. $y = \sqrt{x} + 2$ See the graph in the back of the textbook.

43. $x+2 = \sqrt{8x}$
$(x+2)^2 = \left(\sqrt{8x}\right)^2$
$x^2 + 4x + 4 = 8x$
$x^2 - 4x + 4 = 0$
$(x-2)(x-2) = 0$
$x = 2$

The solution is $x = 2$.

45. Let x = the number
$x - 3 = 2\sqrt{x}$
$(x-3)^2 = \left(2\sqrt{x}\right)^2$
$x^2 - 6x + 9 = 4x$
$x^2 - 10x + 9 = 0$
$(x-9)(x-1) = 0$
$x = 9$

Possible solutions are 1 and 9, only 9 checks.

47. $T = 2$
$2 = \dfrac{11}{7}\sqrt{\dfrac{L}{2}}$
$14 = 11\sqrt{\dfrac{L}{2}}$
$\dfrac{14}{11} = \sqrt{\dfrac{L}{2}}$
$\left(\dfrac{14}{11}\right)^2 = \left(\sqrt{\dfrac{L}{2}}\right)^2$
$\dfrac{196}{121} = \dfrac{L}{2}$
$L = \dfrac{392}{121} \approx 3.24$ feet

49. $\dfrac{x^2-x-6}{x^2-9} = \dfrac{(x-3)(x+2)}{(x+3)(x-3)} = \dfrac{x+2}{x+3}$

51. $\dfrac{x^2-25}{x+4} \cdot \dfrac{2x+8}{x^2-9x+20} = \dfrac{(x+5)(x-5)}{x+4} \cdot \dfrac{2(x+4)}{(x-4)(x-5)} = \dfrac{2(x+5)(x-5)(x+4)}{(x+4)(x-4)(x-5)} = \dfrac{2(x+5)}{x-4}$

© 2000 Harcourt, Inc

53. $\dfrac{x}{x^2-16} + \dfrac{4}{x^2-16} = \dfrac{x+4}{x^2-16} = \dfrac{x+4}{(x+4)(x-4)} = \dfrac{1}{x-4}$

55. $\dfrac{1-\dfrac{25}{x^2}}{1-\dfrac{8}{x}+\dfrac{15}{x^2}} \cdot \dfrac{x^2}{x^2} = \dfrac{x^2-25}{x^2-8x+15} = \dfrac{(x+5)(x-5)}{(x-5)(x-3)} = \dfrac{x+5}{x-3}$

57.
$$\dfrac{x}{x^2-9} - \dfrac{3}{x-3} = \dfrac{1}{x+3}$$
$$(x+3)(x-3)\left(\dfrac{x}{x^2-9} - \dfrac{3}{x-3}\right) = (x+3)(x-3) \cdot \dfrac{1}{x+3}$$
$$x - 3(x+3) = x-3$$
$$x - 3x - 9 = x-3$$
$$-2x - 9 = x - 3$$
$$-3x = 6$$
$$x = -2$$

The solution is $x = -2$.

59. Let t = the time to fill the pool with both pipes open.

$$\dfrac{1}{8} - \dfrac{1}{12} = \dfrac{1}{t}$$
$$24t\left(\dfrac{1}{8} - \dfrac{1}{12}\right) = 24t \cdot \dfrac{1}{t}$$
$$3t - 2t = 24$$
$$t = 24$$

It will take 24 hours to fill the pool with both pipes left open.

61. $y = kx$: $y = 8$, $x = 12$

$$8 = k(12) \Rightarrow k = \dfrac{2}{3}$$
$$y = \dfrac{2x}{3}$$

If $x = 36$
$$y = \dfrac{2(36)}{3} = 24$$

Chapter 8 Review

1. $\sqrt{25} = 5$

3. $\sqrt[3]{-1} = -1$

5. $\sqrt{100x^2y^4} = 10xy^2$

7. $\sqrt{24} = \sqrt{4 \cdot 6} = 2\sqrt{6}$

9. $\sqrt{90x^3y^4} = \sqrt{9x^2y^4 \cdot 10x} = 3xy^2\sqrt{10x}$

11. $3\sqrt{20x^3y} = 3\sqrt{4x^2 \cdot 5xy} = 3 \cdot 2x\sqrt{5xy} = 6x\sqrt{5xy}$

13. $\sqrt{\dfrac{8}{81}} = \dfrac{\sqrt{8}}{\sqrt{81}} = \dfrac{\sqrt{4 \cdot 2}}{9} = \dfrac{2\sqrt{2}}{9}$

15. $\sqrt{\dfrac{49a^2b^2}{16}} = \dfrac{\sqrt{49a^2b^2}}{\sqrt{16}} = \dfrac{7ab}{4}$

17. $\sqrt{\dfrac{40a^2}{121}} = \dfrac{\sqrt{40a^2}}{\sqrt{121}} = \dfrac{\sqrt{4a^2 \cdot 10}}{11} = \dfrac{2a\sqrt{10}}{11}$

19. $\dfrac{3\sqrt{120a^2b^2}}{\sqrt{25}} = \dfrac{3\sqrt{4a^2b^2 \cdot 30}}{\sqrt{25}} = \dfrac{3 \cdot 2ab\sqrt{30}}{5} = \dfrac{6ab\sqrt{30}}{5}$

21. $\dfrac{2}{\sqrt{7}} = \dfrac{2}{\sqrt{7}} \cdot \dfrac{\sqrt{7}}{\sqrt{7}} = \dfrac{2\sqrt{7}}{7}$

23. $\sqrt{\dfrac{5}{48}} = \dfrac{\sqrt{5}}{\sqrt{48}} = \dfrac{\sqrt{5}}{\sqrt{16 \cdot 3}} = \dfrac{\sqrt{5}}{4\sqrt{3}} = \dfrac{\sqrt{5}}{4\sqrt{3}} \cdot \dfrac{\sqrt{3}}{\sqrt{3}} = \dfrac{\sqrt{15}}{4 \cdot 3} = \dfrac{\sqrt{15}}{12}$

25. $\sqrt{\dfrac{32ab^2}{3}} = \dfrac{\sqrt{32ab^2}}{\sqrt{3}} = \dfrac{\sqrt{16b^2 \cdot 2a}}{\sqrt{3}} = \dfrac{4b\sqrt{2a}}{\sqrt{3}} = \dfrac{4b\sqrt{2a}}{\sqrt{3}} \cdot \dfrac{\sqrt{3}}{\sqrt{3}} = \dfrac{4b\sqrt{6a}}{3}$

27. $\dfrac{3}{\sqrt{3}-4} = \dfrac{3}{\sqrt{3}-4} \cdot \dfrac{\sqrt{3}+4}{\sqrt{3}+4} = \dfrac{3\sqrt{3}+12}{(\sqrt{3})^2 - 4^2} = \dfrac{3\sqrt{3}+12}{3-16} = \dfrac{3\sqrt{3}+12}{-13} = \dfrac{-3\sqrt{3}-12}{13}$

29. $\dfrac{3}{\sqrt{5}-\sqrt{2}} = \dfrac{3}{\sqrt{5}-\sqrt{2}} \cdot \dfrac{\sqrt{5}+\sqrt{2}}{\sqrt{5}+\sqrt{2}} = \dfrac{3\sqrt{5}+3\sqrt{2}}{(\sqrt{5})^2 - (\sqrt{2})^2} = \dfrac{3\sqrt{5}+3\sqrt{2}}{5-2} = \dfrac{3\sqrt{5}+3\sqrt{2}}{3} = \dfrac{3(\sqrt{5}+\sqrt{2})}{3} = \sqrt{5}+\sqrt{2}$

31. $\dfrac{\sqrt{5}-\sqrt{2}}{\sqrt{5}+\sqrt{2}} = \dfrac{\sqrt{5}-\sqrt{2}}{\sqrt{5}+\sqrt{2}} \cdot \dfrac{\sqrt{5}-\sqrt{2}}{\sqrt{5}-\sqrt{2}} = \dfrac{5-2\sqrt{10}+2}{(\sqrt{5})^2 - (\sqrt{2})^2} = \dfrac{7-2\sqrt{10}}{5-2} = \dfrac{7-2\sqrt{10}}{3}$

33. $3\sqrt{5} - 7\sqrt{5} = (3-7)\sqrt{5} = -4\sqrt{5}$

35. $-2\sqrt{45} - 5\sqrt{80} + 2\sqrt{20} = -2\sqrt{9 \cdot 5} - 5\sqrt{16 \cdot 5} + 2\sqrt{4 \cdot 5}$
$= -2 \cdot 3\sqrt{5} - 5 \cdot 4\sqrt{5} + 2 \cdot 2\sqrt{5}$
$= -6\sqrt{5} - 20\sqrt{5} + 4\sqrt{5}$
$= (-6 - 20 + 4)\sqrt{5} = -22\sqrt{5}$

37. $\sqrt{40a^3b^2} - a\sqrt{90ab^2} = \sqrt{4a^2b^2 \cdot 10a} - a\sqrt{9b^2 \cdot 10a}$
$= 2ab\sqrt{10a} - a \cdot 3b\sqrt{10a}$
$= 2ab\sqrt{10a} - 3ab\sqrt{10a}$
$= (2-3)ab\sqrt{10a}$
$= -ab\sqrt{10a}$

39. $4\sqrt{2}(\sqrt{3} + \sqrt{5}) = 4\sqrt{2}(\sqrt{3}) + 4\sqrt{2}(\sqrt{5}) = 4\sqrt{6} + 4\sqrt{10}$

41. $(2\sqrt{5} - 4)(\sqrt{5} + 3) = 2\sqrt{5}(\sqrt{5}) + 2\sqrt{5}(3) + (-4)(\sqrt{5}) + (-4)(3)$
$= 2 \cdot 5 + 6\sqrt{5} - 4\sqrt{5} - 12$
$= 10 + 2\sqrt{5} - 12$
$= 2\sqrt{5} - 2$

43. $\sqrt{x-3} = 3$
$(\sqrt{x-3})^2 = 3^2$
$x - 3 = 9$
$x = 12$

45. $5\sqrt{a} = 20$
$(5\sqrt{a})^2 = 20^2$
$25a = 400$
$a = 16$

47. $\sqrt{2x+1} + 10 = 8$
$\sqrt{2x+1} = -2$
$(\sqrt{2x+1})^2 = (-2)^2$
$2x + 1 = 4$
$2x = 3$
$x = \frac{3}{2}$
Does not check. There is no solution.

49. $c = \sqrt{a^2 + b^2}$; Let $a = 1$, $b = \sqrt{2}$, $c = x$
$x = \sqrt{1^2 + (\sqrt{2})^2}$
$x = \sqrt{1 + 2}$
$x = \sqrt{3}$

Cumulative Review: Chapters 1-8

1. $\left(\dfrac{4}{5}\right)\left(\dfrac{5}{4}\right) = \dfrac{20}{20} = 1$

3. $-\left|-\dfrac{1}{2}\right| = -\dfrac{1}{2}$

5. $\dfrac{4^2 - 8^2}{(4-8)^2} = \dfrac{16-64}{(-4)^2} = \dfrac{-48}{16} = -3$

7. $4x - 7x = (4-7)x = -3x$

9. $\dfrac{(b^6)^2 (b^3)^4}{(b^{10})^3} = \dfrac{b^{12} \cdot b^{12}}{b^{30}} = \dfrac{b^{24}}{b^{30}} = b^{24-30} = b^{-6} = \dfrac{1}{b^6}$

11. $\left(\dfrac{1}{2}y + 2\right)\left(\dfrac{1}{2}y - 2\right) = \left(\dfrac{1}{2}y\right)^2 - (2)^2 = \dfrac{1}{4}y^2 - 4$

13. $\dfrac{\frac{1}{a+6} + 3}{\frac{1}{a+6} + 2} \cdot \dfrac{a+6}{a+6} = \dfrac{1 + 3(a+6)}{1 + 2(a+6)} = \dfrac{1 + 3a + 18}{1 + 2a + 12} = \dfrac{3a + 19}{2a + 13}$

15. $\sqrt{\dfrac{90a^2}{169}} = \dfrac{\sqrt{90a^2}}{\sqrt{169}} = \dfrac{\sqrt{9a^2 \cdot 10}}{13} = \dfrac{3a\sqrt{10}}{13}$

17. $\dfrac{7a}{a^2 - 3a - 54} + \dfrac{5}{a-9} = \dfrac{7a}{(a-9)(a+6)} + \dfrac{5}{a-9} \cdot \dfrac{a+6}{a+6}$
$= \dfrac{7a}{(a-9)(a+6)} + \dfrac{5a + 30}{(a-9)(a+6)}$
$= \dfrac{7a + 5a + 30}{(a-9)(a+6)}$
$= \dfrac{12a + 30}{(a-9)(a+6)}$
$= \dfrac{6(2a + 5)}{(a-9)(a+6)}$

19. $3(5x - 1) = 6(2x + 3) - 21$
$15x - 3 = 12x + 18 - 21$
$15x - 3 = 12x - 3$
$3x - 3 = -3$
$3x = 0$
$x = 0$

21. Setting each factor equal to 0 results in $x = 0$, $x = -\dfrac{2}{3}$, or $x = 4$.

23. $16a\left(\dfrac{4}{a}\right) = 16a\left(\dfrac{a}{16}\right)$
$64 = a^2$
$a^2 - 64 = 0$
$(a + 8)(a - 8) = 0$
$a = -8, 8$

25. Adding the two equations:
$2x = 4$
$x = 2$
Substituting into the first equation:
$2 + y = 1$
$y = -1$
The solution is $(2, -1)$.

© 2000 Harcourt, Inc

27. Substitute $y = 3x - 1$ into the first equation:
$$x - (3x - 1) = 5$$
$$x - 3x + 1 = 5$$
$$-2x + 1 = 5$$
$$-2x = 4$$
$$x = -2$$

Substitute into the second equation: $y = 3(-2) - 1 = -6 - 1 = -7$

The solution is $(-2, -7)$.

29. See the graph in the back of the textbook.

31. See the graph in the back of the textbook.

33. To find the x-intercept, let $y = 0$.
$$0 = -x + 7$$
$$x = 7$$
To find the y-intercept, let $x = 0$:
$$y = -(0) + 7 = 7$$

35. Using the point-slope formula:
$$y - (-3) = -1(x - 3)$$
$$y + 3 = -x + 3$$
$$y = -x$$

37. $r^2 + r - 20 = (r + 5)(r - 4)$

39. $x^5 - x^4 - 30x^3 = x^3(x^2 - x - 30) = x^3(x - 6)(x + 5)$

41. $\dfrac{(6 \times 10^5)(6 \times 10^{-3})}{9 \times 10^{-4}} = \dfrac{36 \times 10^2}{9 \times 10^{-4}} = 4 \times 10^6$

43. $\dfrac{5}{\sqrt{3}} \cdot \dfrac{\sqrt{3}}{\sqrt{3}} = \dfrac{5\sqrt{3}}{\sqrt{9}} = \dfrac{5\sqrt{3}}{3}$

45. $5\sqrt{63x^2} - x\sqrt{28} = 5\sqrt{9x^2 \cdot 7} - x\sqrt{4 \cdot 7} = 5 \cdot 3x\sqrt{7} - x \cdot 2\sqrt{7} = 15x\sqrt{7} - 2x\sqrt{7} = 13x\sqrt{7}$

47. $\dfrac{x^2 - 9}{x^4 - 81} = \dfrac{1(x^2 - 9)}{(x^2 + 9)(x^2 - 9)} = \dfrac{1}{x^2 + 9}$

49. Let x = one number
Let $2x + 5$ = the other number
$$x + 2x + 5 = 35$$
$$3x + 5 = 35$$
$$3x = 30$$
$$x = 10$$
$$2x + 5 = 25$$
The two numbers are 10 and 25.

Chapter 8 Test

1. $\sqrt{16} = 4$

2. $-\sqrt{36} = -6$

3. The roots are $\sqrt{49} = 7$ and $-\sqrt{49} = -7$.

4. $\sqrt[3]{27} = 3$

5. $\sqrt[3]{-8} = -2$

6. $-\sqrt[4]{81} = -3$

7. $\sqrt{75} = \sqrt{25 \cdot 3} = \sqrt{25} \cdot \sqrt{3} = 5\sqrt{3}$

8. $\sqrt{32} = \sqrt{16 \cdot 2} = \sqrt{16} \cdot \sqrt{2} = 4\sqrt{2}$

9. $\sqrt{\dfrac{2}{3}} = \dfrac{\sqrt{2}}{\sqrt{3}} = \dfrac{\sqrt{2}}{\sqrt{3}} \cdot \dfrac{\sqrt{3}}{\sqrt{3}} = \dfrac{\sqrt{6}}{3}$

10. $\dfrac{1}{\sqrt[3]{4}} = \dfrac{1}{\sqrt[3]{4}} \cdot \dfrac{\sqrt[3]{2}}{\sqrt[3]{2}} = \dfrac{\sqrt[3]{2}}{\sqrt[3]{8}} = \dfrac{\sqrt[3]{2}}{2}$

11. $3\sqrt{50x^2} = 3\sqrt{25x^2 \cdot 2} = 3\sqrt{25x^2}\sqrt{2} = 3 \cdot 5x\sqrt{2} = 15x\sqrt{2}$

12. $\sqrt{\dfrac{12x^2y^3}{5}} = \dfrac{\sqrt{12x^2y^3}}{\sqrt{5}} = \dfrac{\sqrt{4x^2y^2 \cdot 3y}}{\sqrt{5}} = \dfrac{\sqrt{4x^2y^2}\sqrt{3y}}{\sqrt{5}} = \dfrac{2xy\sqrt{3y}}{\sqrt{5}} = \dfrac{2xy\sqrt{3y}}{\sqrt{5}} \cdot \dfrac{\sqrt{5}}{\sqrt{5}} = \dfrac{2xy\sqrt{15y}}{5}$

13. $5\sqrt{12} - 2\sqrt{27} = 5\sqrt{4 \cdot 3} - 2\sqrt{9 \cdot 3} = 5\sqrt{4}\sqrt{3} - 2\sqrt{9}\sqrt{3} = 5(2)\sqrt{3} - 2(3)\sqrt{3} = 10\sqrt{3} - 6\sqrt{3} = 4\sqrt{3}$

14. $2x\sqrt{18} + 5\sqrt{2x^2} = 2x\sqrt{9 \cdot 2} + 5\sqrt{x^2 \cdot 2} = 2x\sqrt{9}\sqrt{2} + 5\sqrt{x^2}\sqrt{2} = 2x(3)\sqrt{2} + 5x\sqrt{2} = 6x\sqrt{2} + 5x\sqrt{2} = 11x\sqrt{2}$

15. $\sqrt{3}(\sqrt{5} - 2) = \sqrt{3}\sqrt{5} - \sqrt{3}(2) = \sqrt{15} - 2\sqrt{3}$

16. $(\sqrt{5} + 7)(\sqrt{5} - 8) = \sqrt{5}\sqrt{5} - 8\sqrt{5} + 7\sqrt{5} - 7(8) = 5 - 8\sqrt{5} + 7\sqrt{5} - 56 = -51 - \sqrt{5}$

17. $(\sqrt{x} + 6)(\sqrt{x} - 6) = (\sqrt{x})^2 - 6^2 = x - 36$

18. $(\sqrt{5} - \sqrt{3})^2 = (\sqrt{5} - \sqrt{3})(\sqrt{5} - \sqrt{3}) = (\sqrt{5})^2 - 2(\sqrt{5}\sqrt{3}) + (\sqrt{3})^2 = 5 - 2\sqrt{15} + 3 = 8 - 2\sqrt{15}$

19. $\dfrac{\sqrt{7} - \sqrt{3}}{\sqrt{7} + \sqrt{3}} = \dfrac{\sqrt{7} - \sqrt{3}}{\sqrt{7} + \sqrt{3}} \cdot \dfrac{\sqrt{7} - \sqrt{3}}{\sqrt{7} - \sqrt{3}} = \dfrac{(\sqrt{7})^2 - 2(\sqrt{7}\sqrt{3}) + (\sqrt{3})^2}{(\sqrt{7})^2 - (\sqrt{3})^2} = \dfrac{7 - 2\sqrt{21} + 3}{7 - 3} = \dfrac{10 - 2\sqrt{21}}{4} = \dfrac{2(5 - \sqrt{21})}{4} = \dfrac{5 - \sqrt{21}}{2}$

20. $\dfrac{\sqrt{x}}{\sqrt{x} + 5} = \dfrac{\sqrt{x}}{\sqrt{x} + 5} \cdot \dfrac{\sqrt{x} - 5}{\sqrt{x} - 5} = \dfrac{\sqrt{x}\sqrt{x} - 5\sqrt{x}}{(\sqrt{x})^2 - 25} = \dfrac{x - 5\sqrt{x}}{x - 25}$

21. $\sqrt{2x+1} + 2 = 7$
$\sqrt{2x+1} = 5$
$\left(\sqrt{2x+1}\right)^2 = 5^2$
$2x + 1 = 25$
$2x = 24$
$x = 12$

22. $\sqrt{3x+1} + 6 = 2$
$\sqrt{3x+1} = -4$
$\left(\sqrt{3x+1}\right)^2 = (-4)^2$
$3x + 1 = 16$
$3x = 15$
$x = 5$
Since $\sqrt{3 \cdot 5 + 1} + 6 = \sqrt{15 + 1} + 6 = \sqrt{16} + 6 = 4 + 6 = 10 \neq 2$, there is no solution to the equation.

23. $\sqrt{2x-3} = x - 3$
$\left(\sqrt{2x-3}\right)^2 = (x-3)^2$
$2x - 3 = x^2 - 6x + 9$
$0 = x^2 - 8x + 12$
$0 = (x-6)(x-2)$
$x = 2, 6$

$x = 2$, does not check, the solution is $x = 6$.

24.   Let x = a number
$x - 4 = 3\sqrt{x}$
$(x-4)^2 = \left(3\sqrt{x}\right)^2$
$(x-4)(x-4) = \left(3\sqrt{x}\right)\left(3\sqrt{x}\right)$
$x^2 - 8x + 16 = 9x$
$x^2 - 17x + 16 = 0$
$(x-16)(x-1) = 0$
$x = 1, 16$
$x = 1$, does not check, the number is 16.

25. $c = \sqrt{a^2 + b^2}$; $a = 1$, $b = \sqrt{5}$, $c = x$
$x = \sqrt{1^2 + \left(\sqrt{5}\right)^2}$
$x = \sqrt{1 + 5}$
$x = \sqrt{6}$

Chapter 9

9.1 More Quadratic Equations

1. $x^2 = 9$
   $x = \pm\sqrt{9}$
   $= \pm 3$

3. $a^2 = 25$
   $a = \pm\sqrt{25}$
   $= \pm 5$

5. $y^2 = 8$
   $y = \pm\sqrt{8}$
   $= \pm\sqrt{4 \cdot 2}$
   $= \pm\sqrt{4}\sqrt{2}$
   $= \pm 2\sqrt{2}$

7. $2x^2 = 100$
   $x^2 = 50$
   $x = \pm\sqrt{25 \cdot 2}$
   $= \pm\sqrt{25}\sqrt{2}$
   $= \pm 5\sqrt{2}$

9. $3a^2 = 54$
   $a^2 = 18$
   $a = \pm\sqrt{18}$
   $= \pm\sqrt{9 \cdot 2}$
   $= \pm\sqrt{9}\sqrt{2}$
   $= \pm 3\sqrt{2}$

11. $(x+2)^2 = 4$
    $x + 2 = \pm\sqrt{4}$
    $x = -2 \pm \sqrt{4}$
    $= -2 \pm 2$
    $x = 0$ or $x = -4$

13. $(x+1)^2 = 25$
    $(x+1) = \pm\sqrt{25}$
    $x + 1 = \pm 5$
    $x = -1 \pm 5$
    $x = 4$ or $x = -6$

15. $(a-5)^2 = 75$
    $a - 5 = \pm\sqrt{75}$
    $a = 5 \pm \sqrt{75}$
    $= 5 \pm \sqrt{25}\sqrt{3}$
    $= 5 \pm 5\sqrt{3}$

17. $(y+1)^2 = 50$
    $y + 1 = \pm\sqrt{50}$
    $y = -1 \pm \sqrt{50}$
    $= -1 \pm \sqrt{25 \cdot 2}$
    $= -1 \pm \sqrt{25}\sqrt{2}$
    $= -1 \pm 5\sqrt{2}$

19. $(2x+1)^2 = 25$
    $2x + 1 = \pm\sqrt{25}$
    $2x = -1 \pm 5$
    $x = \dfrac{-1 \pm 5}{2}$
    $x = 2$ or $x = -3$

© 2000 Harcourt, Inc

21. $(4a-5)^2 = 36$
$(4a-5) = \pm\sqrt{36}$
$4a-5 = \pm 6$
$4a = 5 \pm 6$
$a = \dfrac{5 \pm 6}{4}$
$a = \dfrac{11}{4}$ or $a = -\dfrac{1}{4}$.

23. $(3y-1)^2 = 12$
$3y-1 = \pm\sqrt{12}$
$3y = 1 \pm \sqrt{12}$
$y = \dfrac{1 \pm \sqrt{12}}{3}$
$= \dfrac{1 \pm \sqrt{4}\sqrt{3}}{3}$
$= \dfrac{1 \pm 2\sqrt{3}}{3}$

25. $(6x+2)^2 = 27$
$(6x+2) = \pm\sqrt{27}$
$6x+2 = \pm\sqrt{9}\sqrt{3}$
$6x+2 = \pm 3\sqrt{3}$
$6x = -2 \pm 3\sqrt{3}$
$x = \dfrac{-2 \pm 3\sqrt{3}}{6}$

27. $(3x-9)^2 = 27$
$3x-9 = \pm\sqrt{27}$
$3x = 9 \pm \sqrt{27}$
$x = \dfrac{9 \pm \sqrt{9}\sqrt{3}}{3}$
$= \dfrac{9 \pm 3\sqrt{3}}{3}$
$= 3 \pm \sqrt{3}$

29. $(3x+6)^2 = 45$
$\sqrt{(3x+6)^2} = \pm\sqrt{45}$
$3x+6 = \pm\sqrt{9}\sqrt{5}$
$3x+6 = \pm 3\sqrt{5}$
$3x = -6 \pm 3\sqrt{5}$
$x = \dfrac{-6 \pm 3\sqrt{5}}{3}$
$x = \dfrac{3(-2 \pm \sqrt{5})}{3}$
$= -2 \pm \sqrt{5}$

31. $(2y-4)^2 = 8$
$2y-4 = \pm\sqrt{8}$
$2y = 4 \pm \sqrt{8}$
$y = \dfrac{4 \pm \sqrt{4}\sqrt{2}}{2}$
$= \dfrac{4 \pm 2\sqrt{2}}{2}$
$= 2 \pm \sqrt{2}$

33. $\left(x - \dfrac{2}{3}\right)^2 = \dfrac{25}{9}$
$x - \dfrac{2}{3} = \pm \dfrac{5}{3}$
$= \dfrac{2}{3} \pm \dfrac{5}{3}$
$x = \dfrac{7}{3}$ or $x = -1$

35. $\left(x + \dfrac{1}{2}\right)^2 = \dfrac{7}{4}$
$x + \dfrac{1}{2} = \pm\sqrt{\dfrac{7}{4}}$
$x = -\dfrac{1}{2} \pm \dfrac{\sqrt{7}}{2}$
$= \dfrac{-1 \pm \sqrt{7}}{2}$

Problem Set 9.1

37. $\left(a - \dfrac{4}{5}\right)^2 = \dfrac{12}{25}$

$a - \dfrac{4}{5} = \pm \dfrac{\sqrt{12}}{5}$

$a = \dfrac{4}{5} \pm \dfrac{2\sqrt{3}}{5}$

$= \dfrac{4 \pm 2\sqrt{3}}{5}.$

39. $x^2 + 10x + 25 = 7$

$(x + 5)^2 = 7$

$x + 5 = \pm\sqrt{7}$

$x = -5 \pm \sqrt{7}$

41. $x^2 - 2x + 1 = 9$

$(x - 1)^2 = 9$

$x - 1 = \pm\sqrt{9}$

$x = 1 \pm 3$

$x = 4 \quad$ or $\quad x = -2$

43. $x^2 + 12x + 36 = 8$

$(x + 6)^2 = 8$

$x + 6 = \pm\sqrt{8}$

$x = -6 \pm \sqrt{4}\sqrt{2}$

$= -6 \pm 2\sqrt{2}$

45. When $x = -1 + 5\sqrt{2}$ the equation

$(x + 1)^2 = 50$ becomes

$\left(-1 + 5\sqrt{2} + 1\right)^2 \overset{?}{=} 50$

$\left(5\sqrt{2}\right)^2 \overset{?}{=} 50$

$\left(5\sqrt{2}\right)\left(5\sqrt{2}\right) \overset{?}{=} 50$

$25(2) \overset{?}{=} 50$

$50 = 50 \quad$ A true statement

47. Let $x =$ a number

$(x + 3)^2 = 16$

$x + 3 = \pm\sqrt{16}$

$x = -3 \pm 4$

$x = 1 \text{ or } -7$

The number is -7 or 1

49. Given $A = 100(1 + r)^2$, solve for r

$A = 100(1 + r)^2$

$\dfrac{A}{100} = (1 + r)^2$

$\pm \dfrac{\sqrt{A}}{10} = 1 + r$

$-1 \pm \dfrac{\sqrt{A}}{10} = r$

$-1 + \dfrac{\sqrt{A}}{10} = r \quad$ (r can't be negative)

51. Let $x =$ height. The height divides the base into two 5-foot sections. Apply the Pythagorean Theorem

$x^2 + 5^2 = 10^2$

$x^2 + 25 = 100$

$x^2 = 75$

$x = \pm\sqrt{75}$

$= \pm\sqrt{25}\sqrt{3}$

$x = 5\sqrt{3}$

The height of the triangle is $5\sqrt{3}$

© 2000 Harcourt, Inc

53. The height divides the base into two equal line segments, of 3 feet each.
Apply the Pythagorean Theorem.
$6^2 = x^2 + 3^2$
$36 = x^2 + 9$
$x^2 = 36 - 9$
$x^2 = 27$
$x = \sqrt{27}$
$x = 3\sqrt{3}$ ft
$x \approx 5.20$ feet
The tent is approximately 5.20 feet high which is 5 ft. $0.2(12 \text{ in.}) = 5$ ft. 2 in.
The answer is no.

55. Let $x =$ the height
The height divides the base into two equal line segments, of 4 feet each.
Apply the Pythagorean Theorem.
$x^2 + 4^2 = 5^2$
$x^2 + 16 = 25$
$x^2 = 9$
$x = \pm\sqrt{9}$
$= 3$
The height is 3 feet.

57. $d = \pi \cdot s \cdot c \cdot \left(\dfrac{1}{2} \cdot b\right)^2$

$192 = (3.14) \cdot (3.53) \cdot (6)\left(\dfrac{1}{2} \cdot b\right)^2$

$192 = 66.5052\left(\dfrac{1}{4}\right)b^2$

$b^2 = 11.548$

$b = \pm\sqrt{11.548}$

$= 3.4$ inches   (no negative values allowed)

Problem Set 9.2

59. $c^2 = a^2 + b^2 : a = 6, c = 14$
$14^2 = 6^2 + b^2$
$196 = 36 + b^2$
$b^2 = 160$
$b = \pm\sqrt{160}$
$= 12.65$ feet
(no negative values allowed)

61. (a) $\overline{AC} = \sqrt{\overline{OC}^2 + \overline{OA}^2}$
$= \sqrt{482^2 + 384^2}$
$= 616$ feet

(b) $\overline{OB} = \sqrt{\overline{BD}^2 + \overline{B'D}^2}$
$= \sqrt{768^2 + 768^2}$
$= 543.1$ feet

$\overline{CB} = \sqrt{\overline{OC}^2 + \overline{OB}^2}$
$= \sqrt{482^2 + 543.1^2}$
$= 726$ feet

(c) $V = \frac{1}{3}\overline{OC}\left(\overline{BD}\right)^2$
$= \frac{1}{3}(482)(768)^2$
$= 94,765,056$ cubic feet

63. $(x-5)^2 = x^2 - 10x + 25$

65. $x^2 - 12x + 36 = (x-6)^2$

67. $x^2 + 4x + 4 = (x+2)^2$

69. $\sqrt[3]{8} = \sqrt[3]{2^3} = 2$

71. $\sqrt[4]{16} = \sqrt[4]{2^4} = 2$

9.2 Completing The Square

1. $\left[(1/2)(6)\right]^2 = 9$
$x^2 + 6x + 9 = (x+3)^2$

3. $\left[(1/2)(2)\right]^2 = 1$
$x^2 + 2x + 1 = (x+1)^2$

5. $\left[(1/2)(-8)\right]^2 = 16$
$y^2 - 8y + 16 = (y-4)^2$

7. $\left[(1/2)(-2)\right]^2 = 1$
$y^2 - 2y + 1 = (y-1)^2$

9. $\left[(1/2)(16)\right]^2 = 64$
$x^2 + 16x + 64 = (x+8)^2$

11. $\left[(1/2)(-3)\right]^2 = \frac{9}{4}$
$a^2 - 3a + \frac{9}{4} = \left(a - \frac{3}{2}\right)^2$

© 2000 Harcourt, Inc

13. $\left[(1/2)(-7)\right]^2 = \dfrac{49}{4}$

$x^2 - 7x + \dfrac{49}{4} = \left(x - \dfrac{7}{2}\right)^2$

15. $\left[(1/2)(1)\right]^2 = \dfrac{1}{4}$

$y^2 + y + \dfrac{1}{4} = \left(y + \dfrac{1}{2}\right)^2$

17. $\left[(1/2)\left(-\dfrac{3}{2}\right)\right]^2 = \dfrac{9}{16}$

$x^2 - \dfrac{3}{2}x + \dfrac{9}{16} = \left(x - \dfrac{3}{4}\right)^2$

19. $x^2 + 4x = 12$

$x^2 + 4x + 4 = 12 + 4$

$(x+2)^2 = 16$

$x + 2 = \pm 4$

$x = -2 \pm 4$

$x = -6 \ \ \text{or} \ \ x = 2$

21. $x^2 - 6x = 16$

$x^2 - 6x + 9 = 16 + 9$

$(x-3)^2 = 25$

$x - 3 = \pm 5$

$x = 3 \pm 5$

$x = -2 \ \ \text{or} \ \ x = 8$

23. $a^2 + 2a = 3$

$a^2 + 2a + 1 = 3 + 1$

$(a+1)^2 = 4$

$a + 1 = \pm 2$

$a = -1 \pm 2$

$a = -3 \ \ \text{or} \ \ a = 1$

25. $x^2 - 10x = 0$

$x^2 - 10x + 25 = 0 + 25$

$(x-5)^2 = 25$

$x - 5 = \pm 5$

$x = 5 \pm 5$

$x = 0 \ \ \text{or} \ \ x = 10$

27. $y^2 + 2y = 15$

$y^2 + 2y + 1 = 15 + 1$

$(y+1)^2 = 16$

$y + 1 = \pm 4$

$y = -1 \pm 4$

$y = -5 \ \ \text{or} \ \ y = 3$

29. $x^2 + 4x - 3 = 0$

$x^2 + 4x = 3$

$x^2 + 4x + 4 = 3 + 4$

$(x+2)^2 = 7$

$x + 2 = \pm\sqrt{7}$

$x = -2 \pm \sqrt{7}$

31. $x^2 - 4x = 4$

$x^2 - 4x + 4 = 4 + 4$

$(x-2)^2 = 8$

$x - 2 = \pm\sqrt{8}$

$x = 2 \pm \sqrt{8}$

$x = 2 \pm 2\sqrt{2}$

Problem Set 9.2

33.
$$a^2 = 7a + 8$$
$$a^2 - 7a = 8$$
$$a^2 - 7a + \frac{49}{4} = 8 + \frac{49}{4}$$
$$\left(a - \frac{7}{2}\right)^2 = \frac{81}{4}$$
$$a - \frac{7}{2} = \pm\frac{9}{2}$$
$$a = \frac{7}{2} \pm \frac{9}{2}$$
$$a = -1 \quad \text{or} \quad a = 8$$

35.
$$4x^2 - 8x - 4 = 0$$
$$x^2 + 2x - 1 = 0$$
$$x^2 + 2x = 1$$
$$x^2 + 2x + 1 = 1 + 1$$
$$(x+1)^2 = 2$$
$$x + 1 = \pm\sqrt{2}$$
$$x = -1 \pm \sqrt{2}$$

37.
$$2x^2 + 2x - 4 = 0$$
$$2x^2 + 2x = 4$$
$$x^2 + x = 2$$
$$x^2 + x + \frac{1}{4} = 2 + \frac{1}{4}$$
$$\left(x + \frac{1}{2}\right)^2 = \frac{9}{4}$$
$$x + \frac{1}{2} = \pm\frac{3}{2}$$
$$x = -\frac{1}{2} \pm \frac{3}{2}$$
$$x = -2 \quad \text{or} \quad x = 1$$

39.
$$4x^2 + 8x + 1 = 0$$
$$4x^2 + 8x = -1$$
$$x^2 + 2x = -1/4$$
$$x^2 + 2x + 1 = -1/4 + 1$$
$$(x+1)^2 = 3/4$$
$$x + 1 = \pm\sqrt{\frac{3}{4}}$$
$$x = -1 \pm \frac{\sqrt{3}}{2}$$
$$= \frac{-2 \pm \sqrt{3}}{2}$$

41.
$$2x^2 - 2x = 1$$
$$x^2 - x = \frac{1}{2}$$
$$x^2 - x + \frac{1}{4} = \frac{1}{2} + \frac{1}{4}$$
$$\left(x - \frac{1}{2}\right)^2 = \frac{3}{4}$$
$$x - \frac{1}{2} = \pm\sqrt{\frac{3}{4}}$$
$$x = \frac{1}{2} \pm \frac{\sqrt{3}}{2}$$
$$= \frac{1 \pm \sqrt{3}}{2}$$

43.
$$4a^2 - 4a + 1 = 0$$
$$4a^2 - 4a = -1$$
$$a^2 - a = -1/4$$
$$a^2 - a + 1/4 = -1/4 + 1/4$$
$$(a - 1/2)^2 = 0$$
$$a - 1/2 = 0$$
$$a = 1/2$$

45. $3y^2 - 9y = 2$

$y^2 - 3y = \dfrac{2}{3}$

$y^2 - 3y + \dfrac{9}{4} = \dfrac{2}{3} + \dfrac{9}{4}$

$\left(y - \dfrac{3}{2}\right)^2 = \dfrac{35}{12}$

$y - \dfrac{3}{2} = \pm \dfrac{\sqrt{35}}{2\sqrt{3}}$

$y - \dfrac{3}{2} = \pm \dfrac{\sqrt{35} \cdot \sqrt{3}}{2\sqrt{3} \cdot \sqrt{3}}$

$y = \dfrac{3}{2} \pm \dfrac{\sqrt{105}}{6}$

$y = \dfrac{9 \pm \sqrt{105}}{6}$

47. $c^2 = a^2 + b^2$

$14^2 = (3x)^2 + (4x)^2$

$196 = 9x^2 + 16x^2$

$196 = 25x^2$

$x^2 = \dfrac{196}{25}$

$x = \pm \sqrt{\dfrac{196}{25}}$

$= \dfrac{14}{5}$ (negative values not allowed)

length $= 4\left(\dfrac{14}{5}\right) = 11.2$ inches

width $= 3\left(\dfrac{14}{5}\right) = 8.4$ inches

49. $x^2 - 2x - 1 = 0$

$x^2 - 2x = 1$

$x^2 - 2x + 1 = 1 + 1$

$(x-1)^2 = 2$

$x - 1 = \pm \sqrt{2}$

$x = 1 \pm \sqrt{2}$

$1 + \sqrt{2} = 1 + 1.4 = 2.4$

$1 - \sqrt{2} = 1 - 1.4 = -0.4$

It crosses the x-axis at $x = -0.4$ and $x = 2.4$

51. (a) $\left(-2 + \sqrt{7}\right) + \left(-2 - \sqrt{7}\right) = -4$

(b) $\left(-2 + \sqrt{7}\right)\left(-2 - \sqrt{7}\right) = (-2)^2 - \left(\sqrt{7}\right)^2$

$= 4 - 7 = -3$

53. See the drawing in the back of the textbook.

55. See the drawing in the back of the textbook.

57. If $a = 2$, $2a = 2(2) = 4$

59. If $a = 2$, and $c = -3$, $4ac = 4(2)(-3) = -24$

61. If $a = 2$, $b = 4$, and $c = -3$, $\sqrt{b^2 - 4ac}$

$= \sqrt{4^2 - 4(2)(-3)}$

$= \sqrt{40}$

$= \sqrt{4 \cdot 10}$

$= 2\sqrt{10}$

63. $\sqrt{12} = \sqrt{4 \cdot 3} = \sqrt{4}\sqrt{3} = 2\sqrt{3}$

65. $\sqrt{20x^2y^3} = \sqrt{4x^2y^2 \cdot 5y} = \sqrt{4x^2y^2}\sqrt{5y} = 2xy\sqrt{5y}$

67. $\sqrt{\dfrac{81}{25}} = \dfrac{\sqrt{81}}{\sqrt{25}} = \dfrac{9}{5}$

9.3 The Quadratic Formula

1. $x^2 + 3x + 2 = 0$,
   $a = 1$, $b = 3$, and $c = 2$,
   $x = \dfrac{-b \pm \sqrt{b^2 - 4ac}}{2a}$
   $= \dfrac{-3 \pm \sqrt{(3)^2 - 4(1)(2)}}{2(1)}$
   $= \dfrac{-3 \pm \sqrt{9 - 8}}{2}$
   $= \dfrac{-3 \pm 1}{2}$
   $x = -1$ or $x = -2$

3. $x^2 + 5x + 6 = 0$
   $a = 1$, $b = 5$, and $c = 6$
   $x = \dfrac{-b \pm \sqrt{b^2 - 4ac}}{2a}$
   $x = \dfrac{-(5) \pm \sqrt{(5)^2 - 4(1)(6)}}{2(1)}$
   $= \dfrac{-(5) \pm \sqrt{25 - 24}}{2}$
   $= \dfrac{-5 \pm 1}{2}$
   $x = -2$ or $x = -3$

5. $x^2 + 6x + 9 = 0$
   $a = 1$, $b = 6$, and $c = 9$,
   $x = \dfrac{-b \pm \sqrt{b^2 - 4ac}}{2a}$
   $= \dfrac{-6 \pm \sqrt{(6)^2 - 4(1)(9)}}{2(1)}$
   $= \dfrac{-6 \pm \sqrt{36 - 36}}{2}$
   $= \dfrac{-6}{2}$
   $x = -3$

7. $x^2 + 6x + 7 = 0$
   $a = 1$, $b = 6$, and $c = 7$
   $x = \dfrac{-b \pm \sqrt{b^2 - 4ac}}{2a}$
   $x = \dfrac{-(6) \pm \sqrt{(6)^2 - 4(1)(7)}}{2(1)}$
   $= \dfrac{-(6) \pm \sqrt{36 - 28}}{2}$
   $= \dfrac{-(6) \pm \sqrt{8}}{2}$
   $= \dfrac{-6 \pm 2\sqrt{2}}{2}$
   $= \dfrac{2(-3 \pm \sqrt{2})}{2}$
   $x = 3 \pm \sqrt{2}$

9. $2x^2 + 5x + 3 = 0$
$a = 2$, $b = 5$, and $c = 3$,
$$x = \frac{-b \pm \sqrt{b^2 - 4ac}}{2a}$$
$$= \frac{-5 \pm \sqrt{(5)^2 - 4(2)(3)}}{2(2)}$$
$$= \frac{-5 \pm \sqrt{25 - 24}}{4}$$
$$= \frac{-5 \pm 1}{4}$$
$x = -1$ or $x = -\frac{3}{2}$

11. $4x^2 + 8x + 1 = 0$
$a = 4$, $b = 8$, and $c = 1$
$$x = \frac{-b \pm \sqrt{b^2 - 4ac}}{2a}$$
$$x = \frac{-(8) \pm \sqrt{(8)^2 - 4(4)(1)}}{2(4)}$$
$$= \frac{-(8) \pm \sqrt{64 - 16}}{8}$$
$$= \frac{-(8) \pm \sqrt{48}}{8}$$
$$= \frac{-4 \cdot 2 \pm 4\sqrt{3}}{4 \cdot 2}$$
$$x = \frac{-2 \pm \sqrt{3}}{2}$$

13. $x^2 - 2x + 1 = 0$
$a = 1$, $b = -2$, and $c = 1$,
$$x = \frac{-b \pm \sqrt{b^2 - 4ac}}{2a}$$
$$= \frac{-(-2) \pm \sqrt{(-2)^2 - 4(1)(1)}}{2(1)}$$
$$= \frac{2 \pm \sqrt{4 - 4}}{2}$$
$$= \frac{2}{2}$$
$x = 1$

15. $x^2 - 5x = 7$
$x^2 - 5x - 7 = 0$
$a = 1$, $b = -5$, and $c = -7$
$$x = \frac{-b \pm \sqrt{b^2 - 4ac}}{2a}$$
$$x = \frac{-(-5) \pm \sqrt{(-5)^2 - 4(1)(-7)}}{2(1)}$$
$$= \frac{5 \pm \sqrt{25 + 28}}{2}$$
$$x = \frac{5 \pm \sqrt{53}}{2}$$

17. $6x^2 - x - 2 = 0$
$a = 6$, $b = -1$, and $c = -2$,
$$x = \frac{-b \pm \sqrt{b^2 - 4ac}}{2a}$$
$$= \frac{-(-1) \pm \sqrt{(-1)^2 - 4(6)(-2)}}{2(6)}$$
$$= \frac{1 \pm \sqrt{1 + 48}}{12}$$
$$= \frac{1 \pm 7}{12}$$
$x = -\frac{1}{2}$ or $x = \frac{2}{3}$

19. $(x - 2)(x + 1) = 3$
$x^2 - x - 2 = 3$
$x^2 - x - 5 = 0$
$a = 1$, $b = -1$, and $c = -5$
$$x = \frac{-b \pm \sqrt{b^2 - 4ac}}{2a}$$
$$x = \frac{-(-1) \pm \sqrt{(-1)^2 - 4(1)(-5)}}{2(1)}$$
$$= \frac{1 \pm \sqrt{1 + 20}}{2}$$
$$x = \frac{1 \pm \sqrt{21}}{2}$$

Problem Set 9.3

21. $(2x-3)(x+2) = 1$
$2x^2 + x - 6 = 1$
$2x^2 + x - 7 = 0$
$a = 2, \ b = 1, \text{ and } c = -7,$
$x = \dfrac{-b \pm \sqrt{b^2 - 4ac}}{2a}$
$= \dfrac{-1 \pm \sqrt{(1)^2 - 4(2)(-7)}}{2(2)}$
$= \dfrac{-1 \pm \sqrt{1 + 56}}{4}$
$x = \dfrac{-1 \pm \sqrt{57}}{4}$

23. $2x^2 - 3x = 5$
$2x^2 - 3x - 5 = 0$
$a = 2, b = -3, \text{ and } c = -5$
$x = \dfrac{-b \pm \sqrt{b^2 - 4ac}}{2a}$
$x = \dfrac{-(-3) \pm \sqrt{(-3)^2 - 4(2)(-5)}}{2(2)}$
$= \dfrac{3 \pm \sqrt{9 + 40}}{4}$
$= \dfrac{3 \pm 7}{4}$
$x = \dfrac{5}{2} \ \text{ or } \ x = -1$

25. $2x^2 = -6x + 7$
$2x^2 + 6x - 7 = 0$
$a = 2, \ b = 6, \text{ and } c = -7,$
$x = \dfrac{-b \pm \sqrt{b^2 - 4ac}}{2a}$
$= \dfrac{-6 \pm \sqrt{(6)^2 - 4(2)(-7)}}{2(2)}$
$= \dfrac{-6 \pm \sqrt{36 + 56}}{4}$
$= \dfrac{-6 \pm \sqrt{92}}{4}$
$= \dfrac{-6 \pm \sqrt{4 \cdot 23}}{4}$
$= \dfrac{-6 \pm 2\sqrt{23}}{4}$
$= \dfrac{2(-3 \pm \sqrt{23})}{4}$
$x = \dfrac{-3 \pm \sqrt{23}}{2}$

27. $3x^2 = -4x + 2$
$3x^2 + 4x - 2 = 0$
$a = 3, b = 4, \text{ and } c = -2$
$x = \dfrac{-b \pm \sqrt{b^2 - 4ac}}{2a}$
$x = \dfrac{-(4) \pm \sqrt{(4)^2 - 4(3)(-2)}}{2(3)}$
$= \dfrac{-4 \pm \sqrt{16 + 24}}{6}$
$= \dfrac{-4 \pm \sqrt{40}}{6}$
$= \dfrac{-4 \pm 2\sqrt{10}}{6}$
$= \dfrac{2(-2 \pm \sqrt{10})}{2 \cdot 3}$
$x = \dfrac{-2 \pm \sqrt{10}}{3}$

© 2000 Harcourt, Inc

29. $2x^2 - 5 = 2x$
$2x^2 - 2x - 5 = 0$
$a = 2, \ b = -2, \text{ and } c = -5,$
$x = \dfrac{-b \pm \sqrt{b^2 - 4ac}}{2a}$
$= \dfrac{-(-2) \pm \sqrt{(-2)^2 - 4(2)(-5)}}{2(2)}$
$= \dfrac{2 \pm \sqrt{4 + 40}}{4}$
$= \dfrac{2 \pm \sqrt{4 \cdot 11}}{4}$
$= \dfrac{2 \pm 2\sqrt{11}}{4}$
$= \dfrac{2(1 \pm \sqrt{11})}{4}$
$x = \dfrac{1 \pm \sqrt{11}}{2}$

31. $2x^3 + 3x^2 - 4x = 0$
$x(2x^2 + 3x - 4) = 0$
$x = 0 \ \text{ or } \ 2x^2 + 3x - 4 = 0$
$a = 2, b = 3, \text{ and } c = -4$
$x = \dfrac{-b \pm \sqrt{b^2 - 4ac}}{2a}$
$x = \dfrac{-(3) \pm \sqrt{(3)^2 - 4(2)(-4)}}{2(2)}$
$= \dfrac{-3 \pm \sqrt{9 + 32}}{4}$
$x = \dfrac{-3 \pm \sqrt{41}}{4} \ \text{ or } \ x = 0$

33. $3x^2 - 4x = 0$
$a = 3, \ b = -4, \text{ and } c = 0,$
$x = \dfrac{-b \pm \sqrt{b^2 - 4ac}}{2a}$
$= \dfrac{-(-4) \pm \sqrt{(-4)^2 - 4(3)(0)}}{2(3)}$
$= \dfrac{4 \pm \sqrt{16}}{6}$
$= \dfrac{4 \pm 4}{6}$
$x = 0 \ \text{ or } \ x = \dfrac{4}{3}$

35. $\dfrac{1}{2}x^2 - \dfrac{1}{2}x - \dfrac{1}{6} = 0$
$3x^2 - 3x - 1 = 0$
$a = 3, b = -3, \text{ and } c = -1$
$x = \dfrac{-b \pm \sqrt{b^2 - 4ac}}{2a}$
$x = \dfrac{-(-3) \pm \sqrt{(-3)^2 - 4(3)(-1)}}{2(3)}$
$= \dfrac{3 \pm \sqrt{9 + 12}}{6}$
$x = \dfrac{3 \pm \sqrt{21}}{6}$

37. $56 = 8 + 64t - 16t^2$
$0 = -48 + 64t - 16t^2$
$0 = -16(3 - 4t + t^2)$
$0 = 3 - 4t + t^2$
$0 = (3 - t)(1 - t)$
$t = 3 \ \text{ or } \ t = 1$
It will be 56 feet above the ground at 1 second and 3 seconds.

39. $(2\sqrt{3})(3\sqrt{5}) = (2 \cdot 3)(\sqrt{3} \cdot \sqrt{5}) = 6\sqrt{15}$

Problem Set 9.4

41. $\left(\sqrt{6}+2\right)\left(\sqrt{6}-5\right) = 6 - 5\sqrt{6} + 2\sqrt{6} - 10 = -4 - 3\sqrt{6}$

43. $(\sqrt{7}-\sqrt{2})(\sqrt{7}+\sqrt{2}) = (\sqrt{7})^2 - (\sqrt{2})^2 = 7 - 2 = 5$

45. $\dfrac{2}{3+\sqrt{5}} = \dfrac{2}{3+\sqrt{5}} \cdot \dfrac{3-\sqrt{5}}{3-\sqrt{5}}$

$= \dfrac{6-2\sqrt{5}}{9-5}$

$= \dfrac{6-2\sqrt{5}}{4}$

$= \dfrac{3-\sqrt{5}}{2}$

9.4 Complex Numbers

1. $(3-2i)+3i = 3+(-2i+3i) = 3+i$

3. $(6+2i)-10i = 6+(2i-10i) = 6-8i$

5. $(11+9i)-9i = 11+(9i-9i) = 11$

7. $(3+2i)+(6-i) = 3+2i+6-i = (3+6)+(2i-i) = 9+i$

9. $(5+7i)-(6+8i) = 5+7i-6-8i = (5-6)+(7i-8i) = -1-i$

11. $(9-i)+(2-i) = 9-i+2-i = (9+2)+(-i-i) = 11-2i$

13. $(6+i)-4i-(2-i) = 6+i-4i-2+i = (6-2)+(i-4i+i) = 4-2i$

15. $(6-11i)+3i+(2+i) = 6-11i+3i+2+i = (6+2)+(-11i+3i+i) = 8-7i$

17. $(2+3i)-(6-2i)+(3-i) = 2+3i-6+2i+3-i = (2-6+3)+(3i+2i-i) = -1+4i$

19. $3(2-i) = 3(2)-3(i) = 6-3i$

21. $2i(8-7i) = 2i(8)-2i(7i) = 16i-14i^2 = 16i-14(-1) = 14+16i$

23. $(2+i)(4-i) = 2\cdot 4 + 2(-i) + i(4) + i(-i) = 8-2i+4i-i^2 = 8+2i-(-1) = 9+2i$

© 2000 Harcourt, Inc

25. $(2+i)(3-5i) = 2\cdot 3 + 2(-5i) + i(3) + i(-5i) = 6 - 10i + 3i - 5i^2 = 6 - 7i - 5(-1) = 11 - 7i$

27. $(3+5i)(3-5i) = (3)^2 - (5i)^2 = 9 - (-25) = 34$

29. $(2+i)(2-i) = (2)^2 - (i)^2 = 4 - (-1) = 5$

31. $\dfrac{2}{3-2i} = \left(\dfrac{2}{3-2i}\right)\left(\dfrac{3+2i}{3+2i}\right) = \dfrac{6+4i}{9-4i^2} = \dfrac{6+4i}{9+4} = \dfrac{6+4i}{13}$

33. $\dfrac{-3i}{2+3i} = \dfrac{-3i}{2+3i}\left(\dfrac{2-3i}{2-3i}\right) = \dfrac{-6i+9i^2}{4-9i^2} = \dfrac{-6i-9}{4+9} = \dfrac{-9-6i}{13}$

35. $\dfrac{6i}{3-i} = \left(\dfrac{6i}{3-i}\right)\left(\dfrac{3+i}{3+i}\right) = \dfrac{18i+6i^2}{9-i^2} = \dfrac{-6+18i}{9+1} = \dfrac{-6+18i}{10} = \dfrac{2(-3+9i)}{2(5)} = \dfrac{-3+9i}{5}$

37. $\dfrac{2+i}{2-i} = \dfrac{2+i}{2-i}\cdot\dfrac{2+i}{2+i} = \dfrac{4+2(2i)+i^2}{4-i^2} = \dfrac{4+4i-1}{4-(-1)} = \dfrac{3+4i}{5}$

39. $\dfrac{4+5i}{3-6i} = \left(\dfrac{4+5i}{3-6i}\right)\left(\dfrac{3+6i}{3+6i}\right) = \dfrac{12+24i+15i+30i^2}{9-36i^2} = \dfrac{12+39i-30}{9+36} = \dfrac{-18+39i}{45} = \dfrac{3(-6+13i)}{3(15)} = \dfrac{-6+13i}{15}$

41. $(x+3i)(x-3i) = x^2 - (3i)^2 = x^2 - 9(-1) = x^2 + 9$

43. $\dfrac{1}{i} = \dfrac{1\cdot i}{i\cdot i} = \dfrac{i}{-1} = -i$

Thus the opposite and the reciprocal of i are the same number.

45. $(x-3)^2 = 25$
$x - 3 = \pm 5$
$x = 3 \pm 5$
$x = -2 \text{ or } x = 8$

47. $(2x-6)^2 = 16$
$2x - 6 = \pm 4$
$2x = 6 \pm 4$
$x = \dfrac{6 \pm 4}{2}$
$x = 5 \text{ or } x = 1$

49. $(x+3)^2 = 12$
$x + 3 = \pm\sqrt{12}$
$x + 3 = \pm 2\sqrt{3}$
$x = -3 \pm 2\sqrt{3}$

Problem Set 9.5

51. $\sqrt{\dfrac{1}{2}} = \dfrac{1}{\sqrt{2}} = \dfrac{1}{\sqrt{2}} \dfrac{\sqrt{2}}{\sqrt{2}} = \dfrac{\sqrt{2}}{2}$

53. $\sqrt{\dfrac{8x^2 y^2}{3}} = \dfrac{\sqrt{8x^3 y^3}}{\sqrt{3}} = \dfrac{\sqrt{4x^2 y^2 \cdot 2y}}{\sqrt{3}} = \dfrac{2xy\sqrt{2y}}{\sqrt{3}} = \dfrac{2xy\sqrt{2y}}{\sqrt{3}} \cdot \dfrac{\sqrt{3}}{\sqrt{3}} = \dfrac{2xy\sqrt{6y}}{3}$

55. $\sqrt[3]{\dfrac{1}{4}} = \dfrac{1}{\sqrt[3]{4}} = \dfrac{1}{\sqrt[3]{4}} \dfrac{\sqrt[3]{2}}{\sqrt[3]{2}} = \dfrac{\sqrt[3]{2}}{2}$

9.5 Complex Solutions to Quadratic Equations

1. $\sqrt{-16} = \sqrt{16(-1)} = \sqrt{16}\sqrt{-1} = 4i$

3. $\sqrt{-49} = \sqrt{49(-1)} = \sqrt{49}\sqrt{-1} = 7i$

5. $\sqrt{-6} = \sqrt{6(-1)} = \sqrt{6}\sqrt{-1} = i\sqrt{6}$

7. $\sqrt{-11} = \sqrt{11(-1)} = \sqrt{11}\sqrt{-1} = i\sqrt{11}$

9. $\sqrt{-32} = \sqrt{32(-1)} = \sqrt{32}\sqrt{-1} = 4i\sqrt{2}$

11. $\sqrt{-50} = \sqrt{50(-1)} = \sqrt{25 \cdot 2}\sqrt{-1} = 5i\sqrt{2}$

13. $\sqrt{-8} = \sqrt{8(-1)} = \sqrt{4 \cdot 2}\sqrt{-1} = 2i\sqrt{2}$

15. $\sqrt{-48} = \sqrt{48(-1)} = \sqrt{16 \cdot 3}\sqrt{-1} = 4i\sqrt{3}$

17.
$$x^2 = 2x - 2$$
$$x^2 - 2x + 2 = 0$$
$$a = 1,\ b = -2,\ \text{and}\ c = 2,$$
$$x = \dfrac{-b \pm \sqrt{b^2 - 4ac}}{2a}$$
$$= \dfrac{-(-2) \pm \sqrt{(-2)^2 - 4(1)(2)}}{2(1)}$$
$$= \dfrac{2 \pm \sqrt{4 - 8}}{2}$$
$$= \dfrac{2 \pm \sqrt{-4}}{2}$$
$$= \dfrac{2 \pm 2i}{2}$$
$$= \dfrac{2(1 \pm i)}{2}$$
$$x = 1 \pm i$$

19.
$$x^2 + 4x = -4$$
$$x^2 - 4x + 4 = 0$$
$$(x - 2)^2 = 0$$
$$x - 2 = 0$$
$$x = 2$$

21. $2x^2 + 5x = 12$
$2x^2 + 5x - 12 = 0$
$(2x - 3)(x + 4) = 0$
$2x - 3 = 0$ or $x + 4 = 0$
$x = \dfrac{3}{2} \qquad x = -4$

23. $(x - 2)^2 = -4$
$x - 2 = \pm\sqrt{-4}$
$x - 2 = \pm 2i$
$x = 2 \pm 2i$

25. $\left(x + \dfrac{1}{2}\right)^2 = -\dfrac{9}{4}$
$x + \dfrac{1}{2} = \pm\sqrt{-\dfrac{9}{4}}$
$x + \dfrac{1}{2} = \pm\dfrac{3}{2}i$
$x = -\dfrac{1}{2} \pm \dfrac{3i}{2}$
$x = \dfrac{-1 \pm 3i}{2}$

27. $\left(x - \dfrac{1}{2}\right)^2 = -\dfrac{27}{36}$
$x - \dfrac{1}{2} = \pm\sqrt{-\dfrac{27}{36}}$
$x - \dfrac{1}{2} = \pm\dfrac{3\sqrt{3}}{6}i$
$x = \dfrac{1}{2} \pm \dfrac{3\sqrt{3}}{6}i$
$x = \dfrac{1}{2} \pm \dfrac{\sqrt{3}}{2}i$
$x = \dfrac{1 \pm i\sqrt{3}}{2}$

29. $x^2 + x + 1 = 0$
$a = 1, \ b = 1, \text{ and } c = 1,$
$x = \dfrac{-b \pm \sqrt{b^2 - 4ac}}{2a}$
$= \dfrac{-1 \pm \sqrt{(1)^2 - 4(1)(1)}}{2(1)}$
$= \dfrac{-1 \pm \sqrt{1 - 4}}{2}$
$= \dfrac{-1 \pm \sqrt{-3}}{2}$
$x = \dfrac{-1 \pm i\sqrt{3}}{2}$

31. $x^2 - 5x + 6 = 0$
$(x - 2)(x - 3) = 0$
$x - 2 = 0$ or $x - 3 = 0$
$x = 2 \qquad x = 3$

Problem Set 9.5

33. $\dfrac{1}{2}x^2 + \dfrac{1}{3}x + \dfrac{1}{6} = 0$

$3x^2 + 2x + 1 = 0$

$a = 3,\ b = 2,\ \text{and}\ c = 1,$

$x = \dfrac{-b \pm \sqrt{b^2 - 4ac}}{2a}$

$= \dfrac{-2 \pm \sqrt{(2)^2 - 4(3)(1)}}{2(3)}$

$= \dfrac{-2 \pm \sqrt{-8}}{6}$

$= \dfrac{-2 \pm 2i\sqrt{2}}{6}$

$= \dfrac{2(-1 \pm i\sqrt{2})}{6}$

$x = \dfrac{-1 \pm i\sqrt{2}}{3}$

35. $\dfrac{1}{3}x^2 = -\dfrac{1}{2}x + \dfrac{1}{3}$

$2x^2 = -3x + 2$

$2x^2 + 3x - 2 = 0$

$(2x - 1)(x + 2) = 0$

$2x - 1 = 0 \quad \text{or} \quad x + 2 = 0$

$x = \dfrac{1}{2} \qquad x = -2$

37. $(x+2)(x-3) = 5$

$x^2 - x - 6 = 5$

$x^2 - x - 11 = 0$

$a = 1,\ b = -1,\ \text{and}\ c = -11,$

$x = \dfrac{-b \pm \sqrt{b^2 - 4ac}}{2a}$

$= \dfrac{-(-1) \pm \sqrt{(-1)^2 - 4(1)(-11)}}{2(1)}$

$= \dfrac{1 \pm \sqrt{45}}{2}$

$= \dfrac{1 \pm \sqrt{9 \cdot 5}}{2}$

$x = \dfrac{1 \pm 3\sqrt{5}}{2}$

39. $(x-5)(x-3) = -10$

$x^2 - 8x + 15 = -10$

$x^2 - 8x + 25 = 0$

$a = 1, b = -8,\ \text{and}\ c = 25$

$x = \dfrac{-b \pm \sqrt{b^2 - 4ac}}{2a}$

$x = \dfrac{-(8) \pm \sqrt{(-8)^2 - 4(1)(25)}}{2(1)}$

$= \dfrac{8 \pm \sqrt{-36}}{2}$

$= \dfrac{8 \pm 6i}{2}$

$= \dfrac{2(4 \pm 3i)}{2}$

$x = 4 \pm 3i$

41. $(2x-2)(x-3) = 9$
$2x^2 - 8x + 6 = 9$
$2x^2 - 8x - 3 = 0$
$a = 2, \ b = -8, \text{ and } c = -3,$
$x = \dfrac{-b \pm \sqrt{b^2 - 4ac}}{2a}$
$= \dfrac{-(-8) \pm \sqrt{(-8)^2 - 4(2)(-3)}}{2(2)}$
$= \dfrac{8 \pm \sqrt{88}}{4}$
$= \dfrac{8 \pm 2\sqrt{22}}{4}$
$= \dfrac{2(4 \pm \sqrt{22})}{4}$
$x = \dfrac{4 \pm \sqrt{22}}{2}$

43. $x^2 - 4x + 5 = 0$
$a = 1, \ b = -4, \text{ and } c = 5,$
$x = \dfrac{-b \pm \sqrt{b^2 - 4ac}}{2a}$
$= \dfrac{-(-4) \pm \sqrt{(-4)^2 - 4(1)(5)}}{2(1)}$
$= \dfrac{4 \pm \sqrt{-4}}{2}$
$= \dfrac{4 \pm 2i}{2}$
$= \dfrac{2(2 \pm i)}{2}$
$x = 2 \pm i$
Since there are no real solutions, the graph does not cross the x-axis.

45. $3 - 7i$ because complex roots are conjugates: $3 \pm 7i$.

47. $y = x - 2$
See the graph in the back of the textbook.

49. $2x + 4y = 8$
See the graph in the back of the textbook.

51. $3\sqrt{50} + 2\sqrt{32} = 3\sqrt{25 \cdot 2} + 2\sqrt{16 \cdot 2} = 15\sqrt{2} + 8\sqrt{2} = 23\sqrt{2}$

53. $\sqrt{24} - \sqrt{54} - \sqrt{150} = \sqrt{4 \cdot 6} - \sqrt{9 \cdot 6} - \sqrt{25 \cdot 6} = 2\sqrt{6} - 3\sqrt{6} - 5\sqrt{6} = -6\sqrt{6}$

55. $2\sqrt{27x^2} - x\sqrt{48} = 2\sqrt{9x^2 \cdot 3} - x\sqrt{16 \cdot 3} = 6x\sqrt{3} - 4x\sqrt{3} = 2x\sqrt{3}$

Problem Set 9.6

9.6 Graphing Parabolas

1. $y = x^2 - 4$
   See the graph in the back of the textbook.

3. $y = x^2 + 5$
   See the graph in the back of the textbook.

5. $y = (x+2)^2$
   See the graph in the back of the textbook.

7. $y = (x-3)^2$
   See the graph in the back of the textbook.

9. $y = (x-5)^2$
   See the graph in the back of the textbook.

11. $y = (x+1)^2 - 2$
    See the graph in the back of the textbook.

13. $y = (x+2)^2 - 3$
    See the graph in the back of the textbook.

15. $y = (x-3)^2 + 2$
    See the graph in the back of the textbook.

17. $y = x^2 + 6x + 5$
    $y = (x^2 + 6x + 9) + 5 - 9$
    $y = (x+3)^2 - 4$
    See the graph in the back of the textbook.

19. $y = x^2 - 2x - 3$
    $y = (x^2 - 2x + 1) - 3 - 1$
    $y = (x-1)^2 - 4$
    See the graph in the back of the textbook.

21. $y = 4 - x^2$
    See the table and the graph
    in the back of the textbook.

23. $y = -1 - x^2$
    See the table and the graph
    in the back of the textbook.

25. $y = x + 2$ and $y = x^2$
    See the graph in the back of the textbook.
    The two graphs intersect at $(-1, 1)$ and $(2, 4)$.

27. $y = 2x^2$ and $y = \dfrac{x^2}{2}$
    See the graph in the back of the textbook.

29. $y = x^2 - 0.25$
    (a) See the graph in the back of the textbook.
    (b) The distance from the vertex to the sun
        is 0.25 million $= 250,000$ miles.
    (c) It crossed the x-axis at $y = 0$.
        $0 = x^2 - 0.25$
        $0.25 = x^2$
        $\pm 0.5 = x$
        The distance from the sun is 500,000 miles

31. $\sqrt{49} = 7$

© 2000 Harcourt, Inc

33. $\sqrt{50} = \sqrt{25 \cdot 2} = \sqrt{25}\sqrt{2} = 5\sqrt{2}$

35. $\sqrt{\dfrac{2}{5}} = \dfrac{\sqrt{2}}{\sqrt{5}} \cdot \dfrac{\sqrt{5}}{\sqrt{5}} = \dfrac{\sqrt{10}}{5}$

37. $3\sqrt{12} + 5\sqrt{27} = 3\sqrt{4 \cdot 3} + 5\sqrt{9 \cdot 3} = 6\sqrt{3} + 15\sqrt{3} = 21\sqrt{3}$

39. $(\sqrt{6}+2)(\sqrt{6}-5) = (\sqrt{6})^2 + \sqrt{6}(-5) + 2\sqrt{6} + 2(-5) = 6 - 5\sqrt{6} + 2\sqrt{6} - 10 = -4 - 3\sqrt{6}$

41. $\dfrac{8}{\sqrt{5}-\sqrt{3}} = \dfrac{8(\sqrt{5}+\sqrt{3})}{(\sqrt{5}-\sqrt{3})(\sqrt{5}+\sqrt{3})} = \dfrac{8(\sqrt{5}+\sqrt{3})}{5-3} = \dfrac{8(\sqrt{5}+\sqrt{3})}{2} = 4\sqrt{5} + 4\sqrt{3}$

43. $\sqrt{2x-5} = 3$
   $2x - 5 = 9$
   $2x = 14$
   $x = 7$

Chapter 9 Review

1. $a^2 = 32$
   $a = \pm\sqrt{32}$
   $a = \pm\sqrt{16 \cdot 2}$
   $a = \pm 4\sqrt{2}$

3. $2x^2 = 32$
   $x^2 = 16$
   $x = \pm\sqrt{16}$
   $x = \pm 4$

5. $(x-2)^2 = 81$
   $x - 2 = \pm 9$
   $x = 2 \pm 9$
   $x = -7 \text{ or } x = 11$

7. $(2x+5)^2 = 32$
   $2x + 5 = \pm\sqrt{32}$
   $2x + 5 = \pm 4\sqrt{2}$
   $2x = -5 \pm 4\sqrt{2}$
   $x = \dfrac{-5 \pm 4\sqrt{2}}{2}$

Chapter 9 Review

9. $\left(x-\dfrac{2}{3}\right)^2 = -\dfrac{25}{9}$

$x - \dfrac{2}{3} = \pm\sqrt{-\dfrac{25}{9}}$

$x - \dfrac{2}{3} = \pm\dfrac{5i}{3}$

$x = \dfrac{2}{3} \pm \dfrac{5i}{3}$

$x = \dfrac{2 \pm 5i}{3}$

11. $x^2 - 8x = 4$

$x^2 - 8x + (4)^2 = 4 + (4)^2$

$(x - 4)^2 = 4 + 16$

$x - 4 = \pm\sqrt{20}$

$x - 4 = \pm 2\sqrt{5}$

$x = 4 \pm 2\sqrt{5}$

13. $x^2 + 4x + 3 = 0$

$x^2 + 4x = -3$

$x^2 + 4x + (2)^2 = -3 + (2)^2$

$(x + 2)^2 = -3 + 4$

$(x + 2)^2 = 1$

$x + 2 = \pm 1$

$x = -2 \pm 1$

$x = -3 \;\; \text{or} \;\; x = -1$

15. $a^2 = 5a + 6$

$a^2 - 5a = 6$

$a^2 - 5a + \left(\dfrac{5}{2}\right)^2 = 6 + \left(\dfrac{5}{2}\right)^2$

$\left(a - \dfrac{5}{2}\right)^2 = \dfrac{24}{4} + \dfrac{25}{4}$

$\left(a - \dfrac{5}{2}\right)^2 = \dfrac{49}{4}$

$a - \dfrac{5}{2} = \pm\dfrac{7}{2}$

$a = \dfrac{5}{2} \pm \dfrac{7}{2}$

$a = -1 \;\; \text{or} \;\; a = 6$

17. $3x^2 - 6x - 2 = 0$

$3x^2 - 6x = 2$

$x^2 - 2x = \dfrac{2}{3}$

$x^2 - 2x + (1)^2 = \dfrac{2}{3} + (1)^2$

$(x - 1)^2 = \dfrac{5}{3}$

$x - 1 = \pm\dfrac{\sqrt{5}}{\sqrt{3}} \cdot \dfrac{\sqrt{3}}{\sqrt{3}}$

$x - 1 = \dfrac{\pm\sqrt{15}}{3}$

$x = 1 \pm \dfrac{\sqrt{15}}{3}$

$x = \dfrac{3 \pm \sqrt{15}}{3}$

© 2000 Harcourt, Inc

19.
$$x^2 - 8x + 16 = 0$$
$$a = 1, \ b = -8, \ c = 16$$
$$x = \frac{-b \pm \sqrt{b^2 - 4ac}}{2a}$$
$$= \frac{-(-8) \pm \sqrt{(-8)^2 - 4(1)(16)}}{2(1)}$$
$$= \frac{8 \pm \sqrt{64 - 64}}{2}$$
$$= \frac{8}{2}$$
$$x = 4$$

21.
$$2x^2 = -8x + 5$$
$$2x^2 + 8x - 5 = 0$$
$$a = 2, \ b = 8, \ c = -5$$
$$x = \frac{-b \pm \sqrt{b^2 - 4ac}}{2a}$$
$$= \frac{-8 \pm \sqrt{8^2 - 4(2)(-5)}}{2(2)}$$
$$= \frac{-8 \pm \sqrt{64 + 40}}{4}$$
$$= \frac{-8 \pm \sqrt{104}}{4}$$
$$= \frac{-8 \pm 2\sqrt{26}}{4}$$
$$= \frac{2(-4 \pm \sqrt{26})}{4}$$
$$x = \frac{-4 \pm \sqrt{26}}{2}$$

23.
$$\frac{1}{5}x^2 - \frac{1}{2}x = \frac{3}{10}$$
$$\frac{1}{5}x^2 - \frac{1}{2}x - \frac{3}{10} = 0$$
$$2x^2 - 5x - 3 = 0$$
$$a = 2, \ b = -5, \ c = -3$$
$$x = \frac{-b \pm \sqrt{b^2 - 4ac}}{2a}$$
$$= \frac{-(-5) \pm \sqrt{(-5)^2 - 4(2)(-3)}}{2(2)}$$
$$= \frac{5 \pm \sqrt{25 + 24}}{4}$$
$$= \frac{5 \pm \sqrt{49}}{4}$$
$$= \frac{5 \pm 7}{4}$$
$$x = -\frac{1}{2} \ \text{or} \ x = 3$$

25. $(4 - 3i) + 5i = 4 + (-3i + 5i) = 4 + 2i$

27. $(5 + 6i) + (5 - i) = (5 + 5) + (6i - i) = 10 + 5i$

29. $(3 - 2i) - (3 - i) = 3 - 2i - 3 + i = (3 - 3) + (-2i + i) = -i$

Chapter 9 Review

31. $(3+i) - 5i - (4-i) = 3 + i - 5i - 4 + i = (3-4) + (i - 5i + i) = -1 - 3i$

33. $2(3-i) = 2(3) - 2(i) = 6 - 2i$

35. $4i(6-5i) = 4i(6) - 4i(5i) = 24i - 20i^2 = 24i - 20(-1) = 20 + 24i$

37. $(3-4i)(5+i) = 15 + 3i - 20i - 4i^2 = 15 - 17i + 4 = 19 - 17i$

39. $(4+i)(4-i) = 16 - i^2 = 16 + 1 = 17$

41. $\dfrac{i}{3+i} = \left(\dfrac{i}{3+i}\right)\left(\dfrac{3-i}{3-i}\right) = \dfrac{3i - i^2}{9 - i^2} = \dfrac{3i + 1}{9 + 1} = \dfrac{1 + 3i}{10}$

43. $\dfrac{5}{2+5i} = \left(\dfrac{5}{2+5i}\right)\left(\dfrac{2-5i}{2-5i}\right) = \dfrac{10 - 25i}{4 - 25i^2} = \dfrac{10 - 25i}{4 + 25} = \dfrac{10 - 25i}{29}$

45. $\dfrac{-3i}{3-2i} = \left(\dfrac{-3i}{3-2i}\right)\left(\dfrac{3+2i}{3+2i}\right) = \dfrac{-9i - 6i^2}{9 - 4i^2} = \dfrac{-9i + 6}{9 + 4} = \dfrac{6 - 9i}{13}$

47. $\dfrac{4-5i}{4+5i} = \left(\dfrac{4-5i}{4+5i}\right)\left(\dfrac{4-5i}{4-5i}\right) = \dfrac{16 + 2(4)(-5i) + (5i)^2}{16 - (5i)^2} = \dfrac{16 - 40i - 25}{16 + 25} = \dfrac{-9 - 40i}{41}$

49. $\sqrt{-36} = \sqrt{36(-1)} = \sqrt{36}\sqrt{-1} = 6i$

51. $\sqrt{-17} = \sqrt{17(-1)} = \sqrt{17}\sqrt{-1} = i\sqrt{17}$

53. $\sqrt{-40} = \sqrt{4(-1)(10)} = \sqrt{4}\sqrt{-1}\sqrt{10} = 2i\sqrt{10}$

55. $\sqrt{-200} = \sqrt{100(-1)(2)} = \sqrt{100}\sqrt{-1}\sqrt{2} = 10i\sqrt{2}$

57. $y = x^2 + 2$
See the graph in the back of the textbook.

59. $y = (x+2)^2$
See the graph in the back of the textbook.

61. $y = x^2 + 4x + 7$
$= (x^2 + 4x + 4) + 7 - 4$
$= (x+2)^2 + 3$
See the graph in the back of the textbook.

© 2000 Harcourt, Inc

Cumulative Review: Chapters 1-9

1. $10 - 8 - 11 = 10 + (-8) + (-11) = -9$

3. $-\dfrac{4}{5} \div \dfrac{8}{15} = -\dfrac{4}{5}\left(\dfrac{15}{8}\right) = -\dfrac{60}{40} = -\dfrac{3}{2}$

5. $\dfrac{(3x^5)(20x^3)}{15x^{10}} = \dfrac{60x^8}{15x^{10}} = \dfrac{4}{x^2}$

7. $\sqrt{81} = 9$

9. $\sqrt{\dfrac{200}{81}} = \dfrac{\sqrt{200}}{\sqrt{81}} = \dfrac{\sqrt{100 \cdot 2}}{9} = \dfrac{10\sqrt{2}}{9}$

11. $(3 + 3i) - 7i - (2 + 2i) = 3 + 3i - 7i - 2 - 2i = 1 - 6i$

13. $\begin{array}{r} a^2 - 6a + 7 \\ \underline{a - 2} \\ a^3 - 6a^2 + 7a \\ \underline{-2a^2 + 12a - 14} \\ a^3 - 8a^2 + 19a - 14 \end{array}$

15. $7 - 4(3x + 4) = -9x$
    $7 - 12x - 16 = -9x$
    $-12x - 9 = -9x$
    $-9 = -9x + 12x$
    $-9 = 3x$
    $x = -3$

17. $5x^2 = -15x$
    $5x^2 + 15x = 0$
    $5x(x + 3) = 0$
    $5x = 0$ or $x + 3 = 0$
    $x = 0$ $x = -3$

19. $\sqrt{6x - 2} = 3x - 5$
    $\left(\sqrt{6x - 2}\right)^2 = (3x - 5)^2$
    $6x - 2 = 9x^2 - 30x + 25$
    $0 = 9x^2 - 36x + 27$
    $0 = 9(x^2 - 4x + 3)$
    $0 = x^2 - 4x + 3$
    $0 = (x - 3)(x - 1)$
    $x - 3 = 0$ or $x - 1 = 0$
    $x = 3$ $x = 1$

© 2000 Harcourt, Inc

Cumulative Review: Chapters 1-9

21. $\dfrac{1}{2}x^2 - \dfrac{1}{3}x = -\dfrac{1}{6}$

$6\left(\dfrac{1}{2}x^2 - \dfrac{1}{3}x\right) = 6\left(-\dfrac{1}{6}\right)$

$3x^2 - 2x = -1$

$3x^2 - 2x + 1 = 0$

$a = 3, b = -2,$ and $c = 1$

$x = \dfrac{-b \pm \sqrt{b^2 - 4ac}}{2a}$

$x = \dfrac{-(-2) \pm \sqrt{(-2)^2 - 4(3)(1)}}{2(3)}$

$= \dfrac{2 \pm \sqrt{-8}}{6}$

$= \dfrac{2 \pm 2i\sqrt{2}}{6}$

$= \dfrac{1 \pm i\sqrt{2}}{3}$

23. $5 \le 2x - 1 \le 7$

$-4 \le 2x \le 8$

$-2 \le x \le 4$

See the graph in the back of the textbook.

25. $x - y = 5$

See the graph in the back of the textbook.

27. $y = -3x + 4$

See the graph in the back of the textbook.

29. $y = (x-3)^2 - 2$

See the graph in the back of the textbook.

31. $x = y - 3$

$2x + 3y = 4$

Substitute $x = y - 3$ into the second equation.

$2(y-3) + 3y = 4$

$2y - 6 + 3y = 4$

$5y - 6 = 4$

$5y = 10$

$y = 2$

Substitute into the first equation.

$x = 2 - 3 = -1$

The solurion is $(-1, 2)$

33. $x^2 - 5x - 24 = (x-8)(x+3)$

35. $25 - y^2 = (5+y)(5-y)$

37. $m = \dfrac{8-(-2)}{1-3} = \dfrac{10}{-2} = -5$

39. $6 - (-2) = 6 + 2 = 8$

© 2000 Harcourt, Inc

41. $\dfrac{15a^3b - 10a^2b^2 - 20ab^3}{5ab} = \dfrac{15a^3b}{5ab} - \dfrac{10a^2b^2}{5ab} - \dfrac{20ab^3}{5ab} = 3a^2 - 2ab - 4b^2$

43. $\dfrac{x^2 + 5x - 24}{x+8} = \dfrac{(x+8)(x-3)}{x-8} = x - 3$

45. $\dfrac{-1}{x^2 - 4} - \dfrac{-2}{x^2 - 4x - 12}$

$= \dfrac{-1}{(x+2)(x-2)} - \dfrac{-2}{(x-6)(x+2)}$

$= \dfrac{-1(x-6)}{(x+2)(x-2)(x-6)} - \dfrac{-2(x-2)}{(x+2)(x-2)(x-6)}$

$= \dfrac{-x + 6 + 2x - 4}{(x+2)(x-2)(x-6)}$

$= \dfrac{x+2}{(x+2)(x-2)(x-6)}$

$= \dfrac{1}{(x-2)(x-6)}$

47. $5\sqrt{200} + 9\sqrt{50} = 5\sqrt{100(2)} + 9\sqrt{25(2)} = 50\sqrt{2} + 45\sqrt{2} = 95\sqrt{2}$

49. Solve for y.
$3x - 8y = 24$
$-8y = -3x + 24$
$y = \dfrac{3}{8}x - 3$

Chapter 9 Test

1. $x^2 - 7x - 8 = 0$
   $(x-8)(x+1) = 0$
   $x - 8 = 0$ or $x + 1 = 0$
   $x = 8 \qquad x = -1$

2. $(x-3)^2 = 12$
   $x - 3 = \pm\sqrt{12}$
   $x - 3 = \pm 2\sqrt{3}$
   $x = 3 \pm 2\sqrt{3}$

3. $\left(x - \dfrac{5}{2}\right)^2 = -\dfrac{75}{4}$
   $x - \dfrac{5}{2} = \pm\sqrt{-\dfrac{75}{4}}$
   $x - \dfrac{5}{2} = \pm\dfrac{5i\sqrt{3}}{2}$
   $x = \dfrac{5}{2} \pm \dfrac{5i\sqrt{3}}{2}$
   $x = \dfrac{5 \pm 5i\sqrt{3}}{2}$

4. $\dfrac{1}{3}x^2 = \dfrac{1}{2}x - \dfrac{5}{6}$
   $6\left(\dfrac{1}{3}x^2\right) = 6\left(\dfrac{1}{2}x\right) - 6\left(\dfrac{5}{6}\right)$
   $2x^2 = 3x - 5$
   $2x^2 - 3x + 5 = 0$
   $a = 2, b = -3,$ and $c = 5$
   $x = \dfrac{-b \pm \sqrt{b^2 - 4ac}}{2a}$
   $= \dfrac{-(-3) \pm \sqrt{(-3)^2 - 4(2)(5)}}{2(2)}$
   $= \dfrac{3 \pm \sqrt{-31}}{4}$
   $x = \dfrac{3 \pm i\sqrt{31}}{4}$

5. $3x^2 = -2x + 1$
   $3x^2 + 2x - 1 = 0$
   $(3x - 1)(x + 1) = 0$
   $3x - 1 = 0$ or $x + 1 = 0$
   $x = \dfrac{1}{3} \qquad x = -1$

6. $(x+2)(x-1) = 6$
   $x^2 + x - 2 = 6$
   $x^2 + x - 8 = 0$
   $a = 1, b = 1$ and $c = -8$
   $x = \dfrac{-b \pm \sqrt{b^2 - 4ac}}{2a}$
   $= \dfrac{-1 \pm \sqrt{(1)^2 - 4(1)(-8)}}{2(1)}$
   $= \dfrac{-1 \pm \sqrt{33}}{2}$

7. $9x^2 + 12x + 4 = 0$
   $(3x+2)^2 = 0$
   $3x + 2 = 0$
   $x = -\dfrac{2}{3}$

8. $x^2 - 6x - 6 = 0$
   $x^2 - 6x = 6$
   $x^2 - 6x + 9 = 6 + 9$
   $(x-3)^2 = 15$
   $x - 3 = \pm\sqrt{15}$
   $x = 3 \pm \sqrt{15}$

9. $\sqrt{-9} = \sqrt{9(-1)} = \sqrt{9}\sqrt{-1} = 3i$

10. $\sqrt{-121} = \sqrt{121(-1)} = \sqrt{121}\sqrt{-1} = 11i$

11. $\sqrt{-72} = \sqrt{72(-1)} = \sqrt{36}\sqrt{-1}\sqrt{2} = 6i\sqrt{2}$

12. $\sqrt{-18} = \sqrt{18(-1)} = \sqrt{9}\sqrt{-1}\sqrt{2} = 3i\sqrt{2}$

13. $(3i+1) + (2+5i) = (1+2) + (3i+5i) = 3 + 8i$

14. $(6-2i) - (7-4i) = 6 - 2i - 7 + 4i = (6-7) + (-2i+4i) = -1 + 2i$

15. $(2+i)(2-i) = (2)^2 - (i)^2 = 4 - (-1) = 5$

16. $(3+2i)(1+i) = 3(1) + 3(i) + 2i(1) + 2i(i) = 3 + 3i + 2i + 2i^2 = 3 + 5i + 2(-1) = 1 + 5i$

17. $\dfrac{i}{3-i} = \dfrac{i}{3-i}\left(\dfrac{3+i}{3+i}\right) = \dfrac{3i+i^2}{9-i^2} = \dfrac{3i-1}{9-(-1)} = \dfrac{-1+3i}{10}$

18. $\dfrac{2+i}{2-i} = \dfrac{2+i}{2-i}\left(\dfrac{2+i}{2+i}\right) = \dfrac{4+2(2i)+i^2}{4-i^2} = \dfrac{4+4i-1}{4-(-1)} = \dfrac{3+4i}{5}$

19. $y = x^2 - 4$
    See the graph in the back of the textbook.

20. $y = (x-4)^2$
    See the graph in the back of the textbook.

21. $y = (x+3)^2 - 4$
    See the graph in the back of the textbook.

22. $y = x^2 - 6x + 11$
    $y = (x^2 - 6x + 9) + 11 - 9$
    $y = (x-3)^2 + 2$
    See the graph in the back of the textbook.

© 2000 Harcourt, Inc

Part Two: Math In Practice Problems

Introduction

As indicated in the Preface at the beginning of this manual, printed versions of the problems on the *Math In Practice: An Applied Video Companion* CD-ROM comprise the second part of this manual. These problems have been formatted in such a way to allow room for you to complete the problems and turn them in to your professor. On the next two pages, there are two correlation charts which provide a breakdown of the problems.

Applications by Concept for *Math in Practice: An Applied Video Companion CD-ROM*

Math Concept	Application	Page	Corresponds to
Basics	Flying to Hawaii	A-1	Chapter 1
	Beetle Infestation	A-3	
	Population Growth	A-5	
	Premature Infant Formula	A-7	
	Aeronautics	A-9	
	World War II Bombers	A-11	
Additional Topic	Aeronautics	A-13	Chapter 1
Linear Equations	Air Pollution	A-15	Chapter 2
	Investing and Debt	A-17	
	Falling Objects	A-19	
	Acoustics	A-21	
	Investing and Debt	A-23	
	Investigating Accidents	A-25	
	Optical Lab	A-27	
Linear Equations	Constructing Buildings	A-29	Chapter 3
Linear Equations - 2 Variables	Flying to Hawaii	A-31	Chapter 3
	Lighting a Stage	A-33	
	Investing & Debt	A-35	
	Metal Alloys	A-37	
	World War II Bombers	A-39	
	Acoustics	A-41	
Systems of Equations	Map Evolution	A-43	Chapter 4
	Air Pollution	A-45	
	Metal Alloys	A-47	
	Premature Infant Formula	A-49	
	Living Soil Crust	A-51	
	Electronics	A-53	
Exponents and Roots	Line of Sight	A-55	Chapter 5
	Population Growth	A-57	
	Atlatl	A-59	
	Optical Lab	A-61	
Basics	Constructing Buildings	A-63	Chapter 5
Rational Equations	Electronics	A-65	Chapter 7
	Flying to Hawaii	A-67	
	Map Evolution	A-69	
	Beetle Infestation	A-71	
	Piano Pitch	A-73	
	Lighting a Stage	A-75	
Rational Exponents	Dinosaur Motion	A-77	Chapter 8
Exponents and Roots	Investigating Accidents	A-79	Chapter 8
	Fighting Fires	A-81	
	Pendulum	A-83	
	Pendulum	A-85	
	Dinosaur Motion	A-87	
	Piano Pitch	A-89	
Quadratic Equations	Population Growth	A-91	Chapter 9
	Atlatl	A-93	
	Optical Lab	A-95	
	Fighting Fires	A-97	
	Line of Sight	A-99	
	Electronics	A-101	

The three problems on pages A-103 to A-107 appear on the CD-ROM but do not apply to McKeague's text. Answers for all problems start on page A-109.

Problems by Application Topic

Math in the Environment   Page Numbers (in this manual)

    Beetle Infestation   A-3, A-71
    Air Pollution   A-15, A-45
    Population Growth   A-5, A-57, A-91
    Living Soil Crust   A-51

Math Over Time

    World War II Bombers   A-11, A-39
    Map Evolution   A-43, A-69
    Atlatl   A-59, A-93
    Dinosaur Motion   A-77, A-87
    Premature Infant Formula   A-7, A-49

Math of Matter and Motion

    Pendulum   A-83, A-85
    Electronics   A-53, A-65, A-101
    Metal Alloys   A-37, A-47
    Line of Sight   A-55, A-99
    Falling Objects   A-19
    Optical Lab   A-27, A-61, A-95

Math on Stage

    Lighting a Stage   A-33, A-75
    Piano Pitch   A-73, A-89
    Acoustics   A-21, A-41

Math at Work

    Fighting Fires   A-81, A-97
    Constructing Buildings   A-29, A-63
    Investigating Accidents   A-25, A-79
    Investing & Debt   A-17, A-23, A-35
    Flying to Hawaii   A-1, A-31, A-67
    Aeronautics   A-9, A-13

The three applications on pages A-103 to A-107 appear on the CD-ROM but do not apply to McKeague's text.

Date _____ Name _____
Course _____ Section _____ Student Number _____

Math Application by Topic: Flying to Hawaii
Math Application by Concept: Basics

There are four time zones separating Salt Lake City, Utah and Honolulu, Hawaii. When it is 3:00 in the afternoon, in Salt Lake City, it is 11:00 in the morning in Honolulu. When Phyllis Upchurch boarded a plane at the Salt Lake City airport at 7:30 A.M. and piloted it west to Hawaii, the trip took 7 hours 40 minutes. What was the local time when she arrived in Honolulu?

© 2000 Harcourt, Inc.

Date _____ Name _____
Course _____   Section _____   Student Number _____

Math Application by Topic: Beetle Infestation
Math Application by Concept: Basics

The video indicates the costs involved with beetle infestation of forests. One cost is the loss of douglas-fir trees. Large douglas-fir trees are the most often attacked because they provide the best conditions for beetles to complete their life cycle.

Wood harvested from timber is measured in board feet. A board foot's value for a douglas-fir is about $0.25 The table below estimates the number of board feet which can be harvested from trees of varying height and diameter.

a. Determine the loss (-) in dollars of a single douglas-fir tree that is 70 feet tall, with a diameter of 19 inches, if the loss of one board foot is -$0.25.
b. One assessment is that the beetle infests 145 million acres in the northwest. If even 5 trees (of similar size to the one calculated above) per acre were infested, what would be a cost estimate of the loss?

Number of Board Feet Harvested
Total Height

Diameter (in inches)	40	50	60	70	80
11	44	57	69	82	95
12	58	72	87	101	115
13	73	89	104	120	136
14	88	105	123	140	158
15	103	122	142	161	181
16	118	139	161	182	204
17	134	157	181	204	227
18	150	175	201	227	253
19	166	194	222	251	279
20	182	213	244	275	306

© 2000 Harcourt, Inc.

Date _____   Name _____
Course _____   Section _____   Student Number _____

Math Application by Topic: Population Growth
Math Application by Concept: Basics

As indicated in the video, one important consideration for planners is identifying dense pockets of population. Drought, lack of employment and climate are all factors which cause groups of people to migrate from one place to another.

The measure of migration used by researchers is "net migration". It is calculated by subtracting the number of people who have moved out of an area from those who have moved into the area: Net Migration = In Migration – Out Migration

Between 1990 to 1991, 346,000 moved into the Northeast region of the U.S. and 932,000 moved out. Use the equation above to determine the Net Migration for the area.

© 2000 Harcourt, Inc.

Date _____ Name _____
Course _____ Section _____ Student Number _____

Math Application by Topic: Premature Infant Formula
Math Application by Concept: Basics

Valerie Johnson explained that the formula for the hyperalumuntation solution used for premature infants was based on the infant's changing weight. The infants weight may change daily or even hourly. The solution described in the video contained: dextrose, protein, and electrolytes A doctor prescribed 2.5 grams of protein kg/day, based on the number of kilograms (kg) of the baby's weight. How many grams of protein would a premature infant require per day who weighed 1.7 kg (almost 3 ¾ pounds)? Round to the nearest tenth of a gram.

© 2000 Harcourt, Inc.

Date _____ Name _____
Course _____ Section _____ Student Number _____

Math Application by Topic: Aeronautics
Math Application by Concept: Basics

As shown in the video, when pilots face unexpected events they have to react within seconds to save the aircraft. The pilot is the one ultimately responsible. For this reason, most late-model aircraft carry systems to assist pilots in dangerous situations such as wind shear. Wind shear is a sudden change in the direction and velocity of the wind and is most dangerous when approaching a runway.

If a pilot is approaching a runway from a height of 2,500 feet and descending at a normal rate of -150 feet per second when a wind shear forces the plane down at an additional rate of -472 feet per second, what would be the plane's altitude after three seconds?

© 2000 Harcourt, Inc.

Date _____ Name _____
Course _____ Section _____ Student Number _____

Math Application by Topic: World War II Bombers
Math Application by Concept: Basics

The B-25 bomber shown in the video was named in honor of Billy Mitchell, a hero of World War I, who tried to persuade the United States government that World War II would be won in the air. He died before the government realized the truth of his argument.

The propellers of the B-25 bomber had a diameter of 12.5 feet. Mechanics felt the plane would lose efficiency if the tips of the propellers rotated faster than the speed of sound (about 761 mph) so they were constantly calculating the appropriate revolutions per minute to rotate the propellers as fast as possible, but under the speed of sound.

a. Determine how fast the tip of each propeller was moving (in feet per minute) when rotating at 1700 revolutions per minute (rpm). Using the diameter of the propeller, determine the circumference of the propeller's rotation ($C = \pi d$). Next multiply by the number of revolutions per minute.
b. Convert your answer (in feet per minute) to calculate how many miles per hour the propellers were rotating. Round to the nearest mile per hour. At 1700 rpm were the propeller tips moving more or less than the speed of sound?

© 2000 Harcourt, Inc.

Date _____ Name _____
Course _____ Section _____ Student Number _____

Math Application by Topic: Aeronautics
Math Application by Concept: Sequences and Series

A pilot checks weather conditions with the National Weather Service before taking off and finds that the air temperature is dropping 4.7°F for every 1000 feet above the surface of the earth. (The higher the flight, the colder the air.)

a. If the air temperature is 38°F when the plane reaches 12,000 feet, write a sequence of numbers that gives the air temperature every 1000 feet as the plane climbs from 12,000 to 18,000 feet. Is this an arithmetic or geometric sequence?
b. Extend the sequence to determine approximately at what altitude the air temperature would fall below 0°F ?

Date _____   Name _____
Course _____   Section _____   Student Number _____

Math Application by Topic: Air Pollution
Math Application by Concept: Linear Equations

The video demonstrated the dramatic increase in carbon dioxide and pollutants in the air over Los Angeles from the 1950s to the 1980s. One automobile can deposit 7,200 pounds of carbon dioxide in the air each year.

Fortunately, carbon dioxide is also necessary for the growth of all plants. Thus, people living within Los Angeles have been urged to plant trees. Scientists have shown that the rate of carbon dioxide consumed annually by mature trees will increase linearly at a rate of 24 lb. per mature tree.

a. Write a linear equation showing the relationship between carbon dioxide consumed and the number of trees planted. Let C equal the total number of pounds of carbon dioxide consumed by the mature trees and T equal the number of trees.
b. Determine how many mature trees would be necessary to consume the carbon dioxide emitted by the automobile described.

© 2000 Harcourt, Inc.

Date _____   Name _____
Course _____   Section _____   Student Number _____

Math Application by Topic: Investing and Debt
Math Application by Concept: Linear Equations

As indicated by the banker, an individual's debt ratio should not exceed 35%. Debt ratio is determined by the total amount of debts divided by an individual's gross income (income before taxes and deductions). Debt ratio = $\dfrac{\text{Debts}}{\text{Gross Income}}$. Amanda is a full-time student, who works part time while attending school. She has exceeded the maximum recommended debt ratio of 35%. If her gross income is $975 for the month, and her debt ratio is 37%,

a. What is the monthly debt she carries?
b. What increase in her gross income must she earn to bring her debt ratio back to 35%?

Date _____ Name _____
Course _____ Section _____ Student Number _____

Math Application by Topic: Falling Objects
Math Application by Concept: Linear Equations

The video showed a bull rider being thrown. He landed on his head and shoulders. We can determine the speed at which the rider hit the ground by using the linear equation described: $s = v + gt$, where s is the speed of the fall, v is the initial velocity that the rider left the bull's back, g is the pull of gravity, and t is the time it took the rider to hit the ground.

Determine what the cowboy's falling speed was when he hit the ground, if his initial downward velocity was 6 feet per second and it took him 0.75 seconds to hit the ground. The pull of gravity is 32 ft/sec^2.

© 2000 Harcourt, Inc.

Date _____ Name _____
Course _____ Section _____ Student Number _____

Math Application by Topic: Acoustics
Math Application by Concept: Linear Equations

As discussed in the video, the time it takes sound to travel can change the way an audience hears a presentation. In large cathedrals a choir director may need to stand half way between the choir at the front of the chapel and the organ at the back in order to conduct the voices and music together.

The length of the main nave at Westminster Abbey is 530 feet. Sounds travels through air at about 1,116 feet per second. (Sound travels somewhat faster on a hot day.)

On a cool day in the fall, how long would it take music played at the back of the chapel to be heard by the choir at the front? Use the relationship Distance = (Rate)(Time). Round your answer to the nearest hundredth of a second.

© 2000 Harcourt, Inc.

Date _____ Name _____
Course _____ Section _____ Student Number _____

Math Application by Topic: Investing and Debt
Math Application by Concept: Linear Equations

The investment banker shown in the video indicated that owning stock in a company may be unpredictable. This is because a company's worth is not measured by how many buildings it owns or its inventory but rather by how much it is worth on the stock market. The market value of a company is called market capitalization or market cap. The equation to calculate market cap of a company is:

Market Cap = (price per share) × (total number of shares outstanding)

The Excelsior-Henderson Motorcycle Manufacturing Company had 13.6 million shares outstanding on the day that it announced it was restructuring its business operations. At the opening of the stock market day, the price per share was $3.00. At the lowest point of the day the price per share had dropped to $2.25 per share. At the end of the trading day, the price per share was $2.375.

a. Use the Market Cap equation to determine the difference in the market value from the company's highest value to its lowest value during the day.
b. If a group of investors owned 12% of the company's stock, what was the market value of their stock at the end of the day?

© 2000 Harcourt, Inc.

Date _____ Name _____
Course _____ Section _____ Student Number _____

Math Application by Topic: Investigating Accidents
Math Application by Concept: Linear Equations

The weight and speed of a car determines the harm that is done to another vehicle at the point of a collision. That is one of the reasons that insurance companies are trying to require sport utility vehicles to pay higher premiums. The damage they inflict on smaller cars is significant.

An estimate of the weight of each car can be calculated with the formula: $W = \dfrac{APN}{2000}$, where W is the vehicle's weight in tons, A is the average tire contact with a hard surface in square inches, P is the air pressure in the tires in pounds per square inch (psi or lb/in^2), and N is the number of tires.

a. In the video two cars were involved in the accident. What is the approximate weight of the Dodge if the average tire contact area is a rectangle 8.0 inches by 10.0 inches, and the air pressure is 38 psi? Round to the nearest tenth.
b. What is the approximate weight of the smaller car if the average tire contact area is a rectangle that is 6.0 inches by 5.0 inches and the tire pressure is 29 psi? Round to the nearest tenth.

© 2000 Harcourt, Inc.

Date _____ Name _____
Course _____ Section _____ Student Number _____

Math Application by Topic: Optical Lab
Math Application by Concept: Linear Equations

The optical technician in the video stated that he uses math constantly in his job, especially when he sizes a lens for an individual. The point where a person looks through the lens needs to be placed correctly in the frame or else blurred vision will occur.

Technicians measure the width of the lens and the width of the frame's bridge. Then they factor in the distance from the center of a client's pupil to the center of the nose. They then work through an algebraic expression. If the value of this expression is less than 5, the glasses are too small. If the value is much greater than 8, the customer will have blurred vision.

The equation for optimizing vision is: $5 \leq \dfrac{(D+B) - 2p}{2} + .5 \leq 8$, where D is the lens width, B is the bridge width and p is the distance from the center of the pupil to center of nose.

A customer came into the optical center to purchase a pair of glasses. The distance from the center of the pupil to the center of the individual's nose was 30 millimeters (mm). The width of the lens was 50 mm. And the width of the bridge was 18 mm.
Would the glasses fit the customer's face correctly?

© 2000 Harcourt, Inc.

Date _____ Name _____
Course _____   Section _____ Student Number _____

Math Application by Topic: Constructing Buildings
Math Application by Concept: Linear Equations

The video describes the components of a roof. Because local building codes dictate the slope of the roof required, builders must be accurate in constructing the pitch.

What is the slope or pitch of a roof that rises 3 feet for every 5 feet of run? The equation $5y = 3x$ may be used to describe the relationship between the amount of rise (y feet) that corresponds with the run (x feet). The slope of the equation is the slope of the roof.

Place the equation in slope-intercept form to determine the slope of the roof. (Solve for y).

© 2000 Harcourt, Inc.

Date _____ Name _____
Course _____ Section _____ Student Number _____

Math Application by Topic: Flying to Hawaii
Math Application by Concept: Linear Equations-2 Variables

Ocean Wide Airlines is flying to LAX from Hawaii. The airplane has an altitude of 30,000 feet (approximately 5.7 miles), and is 65 miles from the airport when it begins its decent into LAX. The equation for the airplane's flight path during its decent is given by $y = -\frac{57}{650}x$. Use the equation to complete the table below. Round your answers to the nearest tenth of a mile. Then use your results to graph the equation. (The graph will appear in the third quadrant only because we are giving distance to the airport in negative numbers.)

Distance to Airport (miles)	Altitude (miles)
-65	
-55	
-45	
-35	
-25	
-15	
-5	

A-31

© 2000 Harcourt, Inc.

Date _____ Name _____
Course _____ Section _____ Student Number _____

Math Application by Topic: Lighting a Stage
Math Application by Concept: Linear Equations-2 variables

The design and structure of a concert hall must be considered when lighting a stage production. When light comes into contact with impenetrable objects, such as costume fabric or parts of the set, it is reflected or absorbed, but not transmitted.

If we let R represent the percent of light reflected and A the percent of light absorbed by a surface, then the relationship between R and A is $R + A = 100$. Graph this equation on a coordinate system where the horizontal axis is the A-axis and the vertical axis is the R axis. Find the intercepts first, and limit your graph to the first quadrant.

© 2000 Harcourt, Inc.

Date _____ Name _____
Course _____ Section _____ Student Number _____

Math Application by Topic: Investing and Debt
Math Application by Concept: Linear Equations

As indicated in the video, an individual's debt burden needs to be carefully considered. Students taking out loans to finance their education are advised that the maximum monthly payment on the amount borrowed for student loans should not exceed 8% of their expected monthly starting salary upon graduation. The recommended maximum monthly payment (y) for such a student loan can be described with the formula: $y = 0.08x$, where x is their predicted monthly salary upon graduation.

a. What is the recommended maximum monthly payment for a journalism student if his/her predicted monthly starting salary is $2350 at a suburban paper in Detroit?
b. What is the needed monthly starting salary for a medical student who was expecting to make a student loan payment of $475?

© 2000 Harcourt, Inc.

Date _____ Name _____
Course _____ Section _____ Student Number _____

Math Application by Topic: Metal Alloys
Math Application by Concept: Linear Equations-2 Variables

The video described the mixing of metal alloys. Brasses are alloys which combine copper and zinc. The properties of brass depend primarily upon the proportion of zinc present but can be modified by the introducing of additional elements to further improve specific characteristics.

Alloys with a higher proportion of zinc cost less because zinc itself costs less. Brasses with a higher proportion of copper and a lower proportion of zinc (10-20%) may be used in decorative architecture or scientific equipment. Brasses with medium amounts of zinc (28-37%) combine the properties of good strength and pliability of copper with the corrosion resistance and a cheap metal price of zinc.

Below is a table of brass alloy prices charged by Mountain Metal.

Type of Brass	% Zinc	Cost/pound
Architectural Brass	15%	$2.55
Brass 260	30%	$2.10
Brass 360	39%	$1.83
Navel Brass 464	40%	$1.80

a. Use the paired data in the table above to create a line graph that shows the relationship between the percent of zinc in the brass alloy and the price per pound.
b. Use any two coordinate pairs to develop an equation which would describe the relationship between price and the copper content in any type of brass.

Date _____   Name _____
Course _____   Section _____   Student Number _____

Math Application by Topic: World War II Bombers
Math Application by Concept: Linear Equations-2 Variables

At the beginning of World War II the United States was at a great disadvantage in the air war. Germany had built up their air force prior to the war, but because the Wright brothers owned most patents for the design of aircraft, the US air corps had not been able to progress as far as the European military.

In 1940 the United States only assembled 6,026 aircraft annually. The U.S. government began to feel that the air war was so important that they had to overcome the issue of patents. By 1944 after a shift in priorities and while the aircraft assembly lines were mainly staffed by newly trained female riveters, 96,318 air craft were produced.

In 1939 Germany was already producing 8,300 aircraft annually. Germany steadily improved the quality and quantity of their air force. At the height of the war, the year 1944, the Third Reich produced 39,800 aircraft.

a. Use the points associated with US manufacturing of aircraft (1940, 6026) and (1944, 96318), to develop a linear equation (in slope-intercept form) to represent the growth in U.S. aircraft manufacturing.
b. Use the points associated with Germany's manufacturing of aircraft (1939, 8300) and (1944, 39800), to develop a linear equation (in slope-intercept form) to represent the growth in German aircraft manufacturing.
c. Use these two equations to determine approximately when U.S. aircraft production over took German aircraft production.

© 2000 Harcourt, Inc.

Date _____ Name _____
Course _____ Section _____ Student Number _____

Math Application by Topic: Acoustics
Math Application by Concept: Linear Equations-2 Variables

A concert hall becomes an extension of a musical instrument. The angles of the walls, materials in room, and size of the space all add subtle echoes and warmth to music. In a well designed music hall the sound and tones are rich and vibrant.

In a much more sterile setting, Dr. Pond demonstrated how an electronic keyboard can synthesize time delay and echo. Because the distance sound travels varies directly as the time it travels ($d = kt$), an electronic keyboard can be programmed to take advantage of the variation constant (k).

If the illusion of sound desired by the musician was that of a large concert hall where sound traveled 336 feet in 0.30 seconds, how far would the sound travel in 0.45 seconds?

Date _____ Name _____
Course _____ Section _____ Student Number _____

Math Application by Topic: Map Evolution
Math Application by Concept: Systems of Equations

As illustrated in the video, wave action constantly erodes coastal regions. Beach replenishment projects are costly. The cost of replenishing each mile of coastal beaches year after year, may be written as a function of time in the equation $C = \$6,000,000t$, where t is the continuing number of years that replenishment projects are necessary.

An alternative to sand replenishing projects is to erect barriers such as seawalls and jetties. These projects have a much larger initial cost, but the maintenance cost over time is less. An equation for the cost of building and maintaining a seawall could be written: $C = \$44,000,000 + \$500,000t$, where t is the continuing number of years that seawall maintenance is required.

Use these two equations to write a system of equations. Use graphing, substitution or elimination to determine at what point in time the seawall becomes the more cost efficient method to protect the coastline.

Date _____   Name _____
Course _____   Section _____   Student Number _____

Math Application by Topic: Air Pollution
Math Application by Concept: Systems of Equations

During a one week period in the 1940s, a small town in Pennsylvania had a temperature inversion resulting in extreme air stagnation. Thousands of people experienced acute illness. Symptoms included coughing, sore throat, chest constriction, shortness of breath, eye irritation, nausea and vomiting and many people died. Seventy percent of the people who died had some form of heart or lung disease.

A system of equations can be used to illustrate the deaths which occurred in this community as a result of polluted air: $x - 20 = 0$ and $7x - 10y = 0$, where x is the number of deaths and y represents the number of deaths who also had some form of heart and lung disease. Solve the system of equations.

© 2000 Harcourt, Inc.

Date _____ Name _____
Course _____ Section _____ Student Number _____

Math Application by Topic: Metal Alloys
Math Application by Concept: Systems of Equations

The development of materials specific to the need of a product is an important part of manufacturing. Metal workers solve systems of equations when forming metal alloys. The video described a 7-ingredient recipe for a new steel consisting of chromium, cobalt, and nickel and other substances.

The Walter Rolling Mills of Oklahoma developed a copper alloy consisting of copper, silver, magnesium and phosphorus. The alloy is very expensive to manufacture and an exact amount of each alloy must be produced to come within the dollar amount bid for the project. Using what is on hand at the mill, the team must create 350 pounds of a metal alloy that is 75% copper. In storage is an alloy which is 82% copper and another alloy is 60% copper. A system of equations may be written to determine the amount of each alloy necessary to develop the product. Write the system of equations and determine the amount.

Date _____   Name _____
Course _____   Section _____   Student Number _____

Math Application by Topic: Premature Infant Formula
Math Application by Concept: Systems of Equations

Valerie Johnson described the need to continually change the proportions of the intravenous solutions given to patients. One solution contained dextrose which was manufactured in categories of : 5%, 10%, 50%, 70% Sometimes patients don't fit into any of these manufactured categories of dextrose and a custom solution must be created.

It was determined that a patient needs a 30% dextrose solution. Using stock solutions of 70% and 10%, determine how much of each solution would be needed to make 3 kl of a 30% dextrose solution.

© 2000 Harcourt, Inc.

Date _____ Name _____
Course _____ Section _____ Student Number _____

Math Application by Topic: Living Soil Crust
Math Application by Concept: Systems of Equations

Lichens are an alliance between fungi and algae. Dr. Rosemary Pendleton described how a living crust of lichens keeps soil from eroding. She indicated that a dry inoculum was needed to rehabilitate damaged soil crust.

Studies are being conducted to find the optimal dry algae inoculum to spread on the soil to promote lichen growth. One study required a 48% algae solution. A lab assistant had produced both a 60% algae solution and a 30% solution. How much of each would be needed to produce 25 gallons of the proposed 48% solution?

© 2000 Harcourt, Inc.

Date _____ Name _____
Course _____ Section _____ Student Number _____

Math Application by Topic: Electronics
Math Application by Concept: Systems of Equations

Use the explanation in the video and the information below to find the amount of electric current flowing through the circuit described.

The following diagram is of an electrical circuit where x, y, and z represent the amount of current (in amperes) flowing across the 5-ohm, 20-ohm, and 10-ohm resistors, respectively. In circuit diagrams resistors are represented by ⋀⋀⋁ and potential differences by ⊣⊢

The system of equations used to find the three currents x, y, and z in the diagram below is:
$x - y - z = 0$
$5x + 20y = 80$
$20y - 10z = 50$

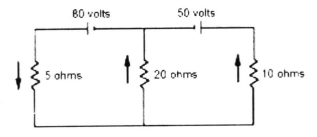

Solve the system for all variables

© 2000 Harcourt, Inc.

Date _____   Name _____
Course _____   Section _____   Student Number _____

Math Application by Topic: Line of Sight
Math Application by Concept: Exponents and Roots

The video discussed how large numbers can be written in scientific notation. Just as the distance from earth to the sun is a large number, the same can be said about the distances between the sun and each of the other planets.

The following table shows the mean distance each planet is from the sun. (The mean distance is listed because planet orbits are not perfectly circular.) Distances are shown in kilometers. Express each of the following distances in scientific notation.

Planet	Mean Distance from Sun (in kilometers)
Mercury	58,000,000
Venus	108,100,000
Earth	150,000,000
Mars	227,840,000
Jupiter	778,100,000
Saturn	1,427,000,000
Uranus	2,870,300,000
Neptune	4,500,000,000
Pluto	5,913,000,000

© 2000 Harcourt, Inc.

Date _____   Name _____
Course _____   Section _____   Student Number _____

Math Application by Topic: Population Growth
Math Application by Concept: Exponents and Roots

The video indicated that only the roughest of estimates exist for the world population. In fact, equation models have only been found to be good for the short term estimates. Each time the U.S. census is counted, adjustments have to be made. One population model for the United States is based on an exponential equation. Using an initial value of 250 million people, the equation predicts a 0.155 percent increase in the population per year.

The population p at year t is computed as $p = 2.5 \times 10^8 (1.00155)^t$. Calculate the predicted population in 20 years ($t = 20$).

© 2000 Harcourt, Inc.

Date _____   Name _____
Course _____   Section _____   Student Number _____

Math Application by Topic: Atlatl
Math Application by Concept: Exponents and Roots

In the video, scientists were studying the additional amount of force and energy which the ancient device gave to a hunting spear. Since the atlatl multiplied the velocity of the spear when thrown, the energy force when the target was hit was greater than that when the spear was thrown by hand.

One scientist in the video made the statement that 15 times the spear's speed gave over 200 times the energy. Use the equation, $I = mr^2$, for calculating the Moment of Inertia (energy) to determine whether the statement was correct. The Moment of Inertia (energy) is I, m is the mass of the spear, and r is the velocity of the spear thrown without the atlatl ($15r$ would be the spear thrown with the atlatl.)

© 2000 Harcourt, Inc.

Date _____ Name _____
Course _____ Section _____ Student Number _____

Math Application by Topic: Optical Lab
Math Application by Concept: Exponents and Roots

The formula $f = \dfrac{ab}{a+b}$ is used in optics to find the focal length of a lens. For an optical lab technician, like that shown in the video, the formula might be used to determine how far the lens should be from the retina of the eye for correct vision. Show that the formula $f = (a^{-1} + b^{-1})^{-1}$ is equivalent to the preceding formula by rewriting it without the negative exponents and then simplifying the results.

Date _____ Name _____
Course _____ Section _____ Student Number _____

Math Application by Topic: Constructing Buildings
Math Application by Concept: Basics

In the video, the construction worker was standing on a 12-foot ladder, which leaned against the house under construction. The base of the ladder was 3 feet from the base of the house. How high did the ladder reach along the side of the building? Round to the nearest tenth of a foot.

© 2000 Harcourt, Inc.

Date _____ Name _____
Course _____ Section _____ Student Number _____

Math Application by Topic: Electronics
Math Application by Concept: Rational Equations

The electronics technician in the video explained the function of a resistor and demonstrated what form parallel resistors took in a circuit. The total resistance of three resistors R_1, R_2, and R_3 in parallel can be given by the equation: $\dfrac{1}{\dfrac{1}{R_1}+\dfrac{1}{R_2}+\dfrac{1}{R_3}}$. Simplify the complex fraction.

© 2000 Harcourt, Inc.

Date _____   Name _____
Course _____   Section _____   Student Number _____

Math Application by Topic: Flying to Hawaii
Math Application by Concept: Rational Equations

As indicated by the pilot in the video, air currents moving east to west affect the speed of aircraft as they travel between the mainland and Hawaii. The average speed for a round trip is given by the complex fraction: Average Speed = $\dfrac{2}{\dfrac{1}{v_1} + \dfrac{1}{v_2}}$, where v_1 is the average speed against an air current west to Honolulu and v_2 is the average speed with an air current on the return trip east. Find the average speed, to the nearest mile per hour, for a round trip if heading toward Hawaii the speed was 270 mph while the plane's speed was 320 mph on the return flight.

© 2000 Harcourt, Inc.

Date _____ Name _____
Course _____ Section _____ Student Number _____

Math Application by Topic: Map Evolution
Math Application by Concept: Rational Equations

To compare the shorelines of past centuries to the current shoreline, a procedure similar to that described in the video is used. The actual distance between points on the historic map must be calculated. The scale on one map drawn in the late 1600s indicates that every 1.3 inches on the map corresponds to an actual distance of 65 miles. Two river outlets were 4 inches apart on the map. Use a proportion to determine the actual distance between the two points at the time the map was drawn.

Date _____ Name _____
Course _____ Section _____ Student Number _____

Math Application by Topic: Beetle Infestation
Math Application by Concept: Rational Equations

In 1999 a study was conducted to test the use of a beetle "scent" called a pheromone to fight a pine beetle infestation in the Idaho Panhandle National Forest. A pheromone is a blend of chemicals naturally produced by beetles. The chemical "scent" tricks insects into thinking the tree is already overrun by beetles, thus hanging out a "no vacancy" sign.

The economics of placing the packets in the forest areas was a concern. For the study approximately 75 pheromone packets were hung in trees on 2.5 acre test plots. If it took 2 hours for one forestry crew to place the pheromone packets on one test plot and it took 3 hours for a second crew to place packets on similar test plot, how long would it take both crews working together to continue placing the pheromone packets on each of the other 2.5 acre plots?

Date _____ Name _____
Course _____ Section _____ Student Number _____

Math Application by Topic: Piano Pitch
Math Application by Concept: Rational Expressions

As Dr. Pond explained, the faster a piano string vibrates, the higher the pitch. Frequency of vibration is measured in Hertz (Hz). A piano's middle C (a reference pitch) will vibrate at 263 times per second and thus has a frequency of 263 Hz. The A string above middle C has a frequency of 440 Hertz.

The equation in the video which explained this relationship between string length and frequency was $F = \dfrac{k}{S}$, where F is the frequency of vibration (pitch), S is the length of the string, and k is a constant of variation (determined by the type of piano). The frequency F varies inversely with the string length. The length of the middle C string on the Yamaha grand piano shown in the video is 60.7 inches. How long would a string, with the same mass, be to play an A note?

Date _____ Name _____
Course _____ Section _____ Student Number _____

Math Application by Topic: Lighting a Stage
Math Application by Concept: Rational Equations

The video described different types of lighting that are needed for a musical in a concert hall. The illumination from any of the light sources in the theater is inversely proportional to the square of the distance from the light. This dissipation of illumination must be taken into account when designing lighting for the production.

If the illumination of a spot light at the back of the hall (a distance of 130 feet) is 50 lumens on the stage actor, what is the illumination at a distance of 75 feet from the source? Use the equation: $I = \dfrac{k}{d^2}$, where I is the amount of illumination, d is the distance, and k is a constant of variation. Round to the nearest lumen.

© 2000 Harcourt, Inc.

Date _____ Name _____
Course _____ Section _____ Student Number _____

Math Application by Topic: Dinosaur Motion
Math Application by Concept: Rational Equations

The animators of the dinosaurs in the movie Jurassic Park used a computer program to build the dinosaurs on film. As the actors were running, the animated dinosaurs had to look as if they were moving within the same environment. The computer program used mathematical ratios similar to the following: $Dimensionless\ speed = \dfrac{speed}{\sqrt{lg}}$, where l = leg length, g = gravitational acceleration and *speed* is the speed that the actor was running.

If an actor was moving at a *speed* of 15 feet per second, had a leg length l of 3 feet, with gravitational acceleration g of 16 feet per second, what would be the ratio for *Dimensionless speed* that the computer would use to animate the dinosaurs? Round to the nearest tenth.

© 2000 Harcourt, Inc.

Date _____ Name _____
Course _____ Section _____ Student Number _____

Math Application by Topic: Investigating Accidents
Math Application by Concept: Exponents and Roots

The accident that was described in the video was on the street outside the studio. The investigating officer described a method for calculating how fast the vehicle was going when the breaks were first applied using the coefficient of friction of the road surface and the length of the skid marks. The formula, $S = \sqrt{30df}$ was used where S is the speed of vehicle in mph, d is the length of the skid, and f is a coefficient of friction.

Use the formula to determine how fast (to the nearest mile/hour) the driver was going when the breaks were applied. Because of the bad weather conditions, the wet road had a friction coefficient (f) of 0.55 The tire's skid marks measured 117 feet.

Answer: 44 miles per hour

© 2000 Harcourt, Inc.

Date _____   Name _____
Course _____   Section _____   Student Number _____

Math Application by Topic: Fighting Fires
Math Application by Concept: Exponents and Roots

The pressure of a stream of water from a fire hydrant can be calculated from the equation $G = 26.8D^2 P^{1/2}$ or $G = 26.8D^2 \sqrt{P}$, where G is the discharge of water (in gallons per minute); D is the diameter of the outlet (in inches); and P is the water pressure (in pounds per square inch).

A "feather-light" nozzle was advertised by the Cordoma Fire Equipment Co. as a heavy-duty nozzle with a weight of only 10 ounces. A fire fighter questions the ability of such a light nozzle to sustain the appropriate amount of water pressure. The nozzle was tested. The diameter was 1.5 inches and produced a stream of 350 gallons per minute. What was the water pressure at the nozzle? Round to the nearest tenth.

© 2000 Harcourt, Inc.

Date _____   Name _____
Course _____   Section _____   Student Number _____

Math Application by Topic: Pendulum
Math Application by Concept: Exponents and Roots

The Foucault pendulum shown in the video has a large weight attached to the end of a long cable. The cable is attached to the ceiling of the second story of the structure. The length of the wire cable is 30 feet to the bottom of the weighted bob.

a. Use the equation described in the video, $T = 2\pi \dfrac{\sqrt{L}}{\sqrt{32}}$, to determine the cycle of the pendulum. Let T be the time of a cycle in seconds and L the length of the pendulum in feet. Remember that a cycle is a swing in one direction and back again.
b. Rewrite the formula so it may be used to determine the Length (L) of the cable. (Solve for L).

© 2000 Harcourt, Inc.

Date _____ Name _____
Course _____ Section _____ Student Number _____

Math Application by Topic: Pendulum
Math Application by Concept: Exponents and Roots

The antique pendulum clock shown in the video should have a cycle of one second long to keep accurate time. Sometimes when the clock's time is too slow or fast, the length of the pendulum must be adjusted. This can be done by raising or lowering the bob at the bottom thus making the pendulum length shorter or longer.

Using the equation described, $T = 2\pi \dfrac{\sqrt{L}}{\sqrt{32}}$, where T is the time of a cycle and L is the length of the pendulum, how many feet long should the pendulum be to keep accurate time?

Date _____ Name _____
Course _____ Section _____ Student Number _____

Math Application by Topic: Dinosaur Motion
Math Application by Concept: Exponents and Roots

The scientist in the video described how the speed of a dinosaur's movements is determined by the height of the animal's hips and the length of its stride. When two dinosaurs have similar movement, then a ratio for dimensionless speed can be used. This type of equation can predict the speed and movement of similarly structured animals. A comparable equation is used for the animation of dinosaurs in movies such as Jurassic Park to make the dinosaurs and people look as if they were moving within the same scene.

Use the equation: Dimensionless speed $= \dfrac{speed}{\sqrt{lg}}$, where l = leg length, and g = gravitational acceleration to find the speed that a dinosaur moves who has a leg length of 2.8 meters. Let the dimensionless speed ratio be 0.4 Let gravitational acceleration be approximately 10 meters per second. Find the speed the dinosaur moves to the nearest tenth of a meter per second.

© 2000 Harcourt, Inc.

Date _____ Name _____
Course _____ Section _____ Student Number _____

Math Application by Topic: Piano Pitch
Math Application by Concept: Exponents and Roots

Dr. Pond described the variables in tuning a grand piano. The tension on each string was one of the variables which determined the pitch of the note. The greater the tension, the higher the note. The tension of the strings is held by pegs which can be tightened or loosened.

As stated in the video, the pitch (or frequency) of the note varies directly with the square root of the tension of the string. The equation used was: $F = k\sqrt{T}$, where F is the frequency (measured in Hertz), T is the tension on the string (measured in pounds), and k is a constant of variation.

a. Suppose a string is tuned to a frequency of 233 Hertz (an A sharp) and there are 152 pounds of tension on the string. Use the equation of frequency: $F = k\sqrt{T}$ to find the value of the constant of variation (k). Round to the nearest whole number.
b. Considering the relationship between frequency and tension, raising any note an octave (doubling the frequency) would require how much more tension on a string?

Date _____   Name _____
Course _____   Section _____   Student Number _____

Math Application by Topic: Population Growth
Math Application by Concept: Quadratic Equations

In 1829, former President James Madison made a prediction about the growth of the population of the United States. He felt the population growth could be modeled by the equation, $y = .029x^2 - 1.39x + 42$, where y was the population in millions of people x years from 1829.

Use the equation to determine the approximate year President Madison would have predicted that the U.S. population would reach 100,000,000 ($y = 100$).

© 2000 Harcourt, Inc.

Date _____ Name _____
Course _____ Section _____ Student Number _____

Math Application by Topic: Atlatl
Math Application by Concept: Quadratic Equation

The distance an arrow travels is a function of time. The longer an arrow is in the air, the farther it flies. The compound bow shown in the video can shoot an arrow for a long distance. The equation, $S = -16t^2 + 192t$ describes the distance the arrow in the video would travel when shot straight into the air from a compound bow. S is the distance and t is the time involved.

a. What is the distance the arrow would travel in 1.5 seconds?
b. How long would it take the arrow to drop back down to the ground ($S = 0$)?

Date _____ Name _____
Course _____ Section _____ Student Number _____

Math Application by Topic: Optical Lab
Math Application by Concept: Quadratic Equations

Every optical lab can produce products in different amounts of time. There are many factors which can lead to this: skill of the employees, age of the equipment, grade of the materials, methods used, or employee motivation.

One equation which could be used to determine the number of glasses produced in a month by the optical lab in the video is based on the number of employees: $G = 2n^2 - 3n + 5$. Where G is the number of pairs of glasses and n is the number of employees working at the lab. For this equation assume all employees work at the same rate.

Use the equation to determine how many pairs of glasses can be produced by 24 employees working at the lab during a month.

© 2000 Harcourt, Inc.

Date _____ Name _____
Course _____ Section _____ Student Number _____

Math Application by Topic: Fighting Fires
Math Application by Concept: Quadratic Equations

As shown in the video, firefighters have gauges which are placed at the nozzle of the hose to measure the water flow velocity in feet per second. A tight straight stream is used for penetrating a burning structure, while a fog is for mopping-up at the end of a fire.

The equation: $V^2 = 147P$ shows the relationship between the water pressure (P) at the nozzle point and the velocity (V) of the water flow (in feet per second) as it is sprayed from the hose.

Determine the velocity (to the nearest foot per second) of the water discharged from a nozzle if the gauge indicates 52 pounds of pressure per square inch (psi).

© 2000 Harcourt, Inc.

Date _____   Name _____
Course _____   Section _____   Student Number _____

Math Application by Topic: Line of Sight
Math Application by Concept: Quadratic Equations

The video discusses an equation which pilots use to determine the distance to the horizon by using the altitude of the plane as determined by the instrument panel. The equation is $D = 1.2\sqrt{A}$ where D is measured in miles and A is measured in feet.

a. Rewrite this equation to solve for the variable A.
b. Use your rewritten equation to approximate the altitude of a hot air balloon if the pilot can see the buildings of Springfield, Missouri while flying above Branson (40 miles away). Round to the nearest hundred feet.

Date _____ Name _____
Course _____ Section _____ Student Number _____

Math Application by Topic: Electronics
Math Application by Concept: Quadratic Equations

Complex numbers may be applied to electrical circuits. Electrical engineers use the fact that resistance R to electrical flow, flow of the current I, and the voltage V, are related by the formula $V = RI$. (Voltage is measured in volts, resistance in ohms, and current in amperes.)

Find the resistance to electrical flow in a circuit that has a voltage $V = (80 + 20i)$ volts and current $I = (-6 + 2i)$ amps.

© 2000 Harcourt, Inc.

Date _____   Name _____
Course _____   Section _____   Student Number _____

Math Application by Topic: Earthquakes
Math Application by Concept: Additional Topics

The video showed footage of an earthquake in Japan. That earthquake which hit Kobe, Japan in 1992 recorded a magnitude of $M = 7.2$. The earthquake with the greatest magnitude ($M = 8.9$) ever recorded was also in Japan. Because the Richter scale is a logarithmic scale the difference between the two numbers is not very much, but the shock waves they measure are very different. The magnitude M of an earthquake is given by the formula: $M = \log_{10} T$, where T is the relative intensity of the shock wave.

Use the formula, $M = \log_{10} T$, in exponential form, to determine how many times greater the shock wave T was during the great Japanese earthquake ($M = 8.9$) than the Kobe earthquake of 1992 ($M = 7.2$) In solving round T to the nearest 10 million.

A-103

© 2000 Harcourt, Inc.

Date _____ Name _____
Course _____ Section _____ Student Number _____

Math Application by Topic: Earthquakes
Math Application by Concept: Additional Topics

The video indicated that there are approximately 6000 earthquakes worldwide per year, but 5500 of these earthquakes are either too small or remote to be felt. On a Richter Scale the magnitude M of an earthquake is given by the formula: $M = \log_{10} T$, where T is the relative intensity of the shock wave to the smallest shock wave measurable. What would be the magnitude of an earthquake that had a shock wave of $T = 100{,}000$?

Date _____   Name _____
Course _____   Section _____   Student Number _____

Math Application by Topic: Piano Pitch
Math Application by Concept: Additional Topics

Dr. Pond discussed the harmonics of music as it applies to the length, mass and tension of piano strings. The sequence $1, \frac{1}{2}, \frac{1}{3}, \frac{1}{4}, \frac{1}{5}, \ldots$ is a natural harmonic sequence and can be mathematically associated with a vibrating musical string.

a. Find the next term in this sequence.
b. Find the 78th term in this sequence.
c. Write an expression for the term number n in the sequence.

Answers to *Math in Practice; An Applied Video Companion CD-ROM*

Math Application by Topic: Flying to Hawaii
Math Application by Concept: Basics
Answer: 11:10 AM

Math Application by Topic: Beetle Infestation
Math Application by Concept: Basics
Answers:
a.   -$62.75
b.   -$45,493,750,000

Math Application by Topic: Population Growth
Math Application by Concept: Basics
Answer: Loss of 585,000 people (-585,000)

Math Application by Topic: Premature Infant Formula
Math Application by Concept: Basics
Answer: 4.3 grams

Math Application by Topic: Aeronautics
Math Application by Concept: Basics
Answer: 634 feet

Math Application by Topic: World War II Bombers
Math Application by Concept: Basics
Answers:
a.   66,725 feet per minute
b.   Approximately 758 miles per hour which is slightly less than the speed of sound.

Math Application by Topic: Aeronautics
Math Application by Concept: Sequences and Series
Answers:
a.   {38°F, 33.3°F, 28.6°F, 23.9°F, 19.2°F, 14.5°F, 9.8°F}; Arithmetic sequence
b.   Between 20,000 and 21,000 feet.

Math Application by Topic: Air Pollution
Math Application by Concept: Linear Equations
Answers:
a. $C = 24T$
b.   300 trees

Math Application by Topic: Investing and Debt
Math Application by Concept: Linear Equations
Answers:
a. $360.75
b. $55.71

Math Application by Topic: Falling Objects
Math Application by Concept: Linear Equations
Answer: 30 feet per second

Math Application by Topic: Acoustics
Math Application by Concept: Linear Equations
Answer: $T \approx 0.47$ second

Math Application by Topic: Investing and Debt
Math Application by Concept: Linear Equations
Answers:
a.   Difference during day = $10.2 million
b.   Market value of stock at end of day = $3,876,000

Math Application by Topic: Investigating Accidents
Math Application by Concept: Linear Equations
Answers:
a.   6.1 tons
b.   1.7 tons

Math Application by Topic: Optical Lab
Math Application by Concept: Linear Equations
Answer: The glasses would not fit the customer's face correctly.
4.5 is not between 5 and 8.

Math Application by Topic: Constructing Buildings
Math Application by Concept: Linear Equations

Answer: The slope is: $\dfrac{3}{5}$

Math Application by Topic: Flying to Hawaii
Math Application by Concept: Linear Equations-2 Variables
Answer:

Distance to Airport (miles)	Altitude (miles)
-65	5.7
-55	4.8
-45	3.9
-35	3.1
-25	2.2
-15	1.3
-5	0.4

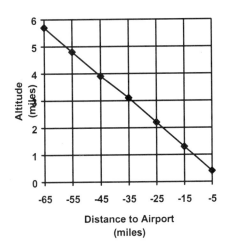

A-111

Math Application by Topic: Lighting a Stage
Math Application by Concept: Linear Equations-2 variables
Answer:

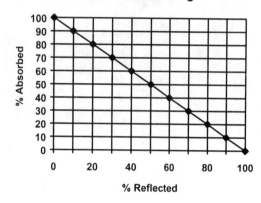

Math Application by Topic: Investing and Debt
Math Application by Concept: Linear Equations
Answers:
a. $188.00
b. $5937.50

Math Application by Topic: Metal Alloys
Math Application by Concept: Linear Equations-2 Variables
Answers:
a.   See graph on next page.

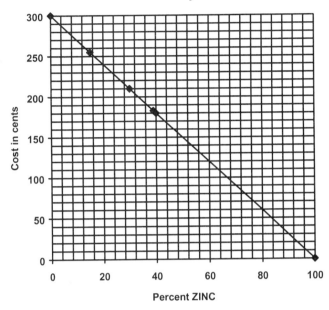

b. $y = -3x + 300$, where x = % of zinc and y = cost in cents

---

Math Application by Topic: World War II Bombers
Math Application by Concept: Linear Equations-2 Variables
Answers: a. $y = 22573x - 43785594$
b. $y = 6300x - 12207400$
c. $x \approx 1940.5$ Thus, about midway through 1940 the US production of airplanes overtook German production.

Math Application by Topic: Acoustics
Math Application by Concept: Linear Equations-2 Variables
Answer: 504 feet

Math Application by Topic: Map Evolution
Math Application by Concept: Systems of Equations
Answer: 8 years

Math Application by Topic: Air Pollution
Math Application by Concept: Systems of Equations
Answer:
$x = 20$ and $y = 14$

Math Application by Topic: Metal Alloys
Math Application by Concept: Systems of Equations
Answer:
   Approximately 238.64 pounds of the alloy with 82% copper
   Approximately 111.36 pounds of the alloy with 60% copper

Math Application by Topic: Premature Infant Formula
Math Application by Concept: Systems of Equations
Answer:
   2 kl of 10% dextrose solution
   1 kl of 70% dextrose solution

Math Application by Topic: Living Soil Crust
Math Application by Concept: Systems of Equations
Answer:
   10 gallons of the 30% solution
   15 gallons of the 60% solution

Math Application by Topic: Electronics
Math Application by Concept: Systems of Equations
Answer: $x = 4$, $y = 3$, $z = 1$

Math Application by Topic: Line of Sight
Math Application by Concept: Exponents and Roots
Answer:
   Mercury 5.8×10^7
   Venus 1.081×10^8
   Earth 1.5×10^8
   Mars 2.2784×10^8
   Jupiter 7.781×10^8
   Saturn 1.427×10^9
   Uranus 2.8703×10^9
   Neptune 4.5×10^9
   Pluto 5.913×10^9

Math Application by Topic: Population Growth
Math Application by Concept: Exponents and Roots
Answer: $p \approx 25{,}786{,}500$ people

Math Application by Topic: Atlatl
Math Application by Concept: Exponents and Roots
Answer: The statement was correct. $I = 225mr^2$

Math Application by Topic: Optical Lab
Math Application by Concept: Exponents and Roots
Answer:
$$f = \frac{ab}{a+b}$$

Math Application by Topic: Constructing Buildings
Math Application by Concept: Basics
Answer: 11.6 feet

Math Application by Topic: Electronics
Math Application by Concept: Rational Equations

Answer: $\dfrac{R_1 R_2 R_3}{R_2 R_3 + R_1 R_3 + R_1 R_2}$

Math Application by Topic: Flying to Hawaii
Math Application by Concept: Rational Equations
Answer: Average Speed ≈ 293 mph

Math Application by Topic: Map Evolution
Math Application by Concept: Rational Equations
Answer: 200 miles

Math Application by Topic: Beetle Infestation
Math Application by Concept: Rational Equations
Answer: $\dfrac{6}{5}$ hours or 1 hour 12 minutes.

Math Application by Topic: Piano Pitch
Math Application by Concept: Rational Expressions
Answer: String length for the A note ≈ 36.3 inches

Math Application by Topic: Lighting a Stage
Math Application by Concept: Rational Equations
Answer: 15 lumens

Math Application by Topic: Dinosaur Motion
Math Application by Concept: Rational Equations
Answer: Dimensionless speed ≈ 2.2

Math Application by Topic: Investigating Accidents
Math Application by Concept: Exponents and Roots
Answer: 44 miles per hour

Math Application by Topic: Fighting Fires
Math Application by Concept: Exponents and Roots
Answer: $P \approx 33.6$ psi (pounds per square inch)

Math Application by Topic: Pendulum
Math Application by Concept: Exponents and Roots
Answers:
a. $T \approx 6.1$ seconds. Rounded to the nearest tenth.
b. $L = \dfrac{8T^2}{\pi^2}$

Math Application by Topic: Pendulum
Math Application by Concept: Exponents and Roots
Answer: 0.81 foot. Rounded to the nearest hundredth.

Math Application by Topic: Dinosaur Motion
Math Application by Concept: Exponents and Roots
Answer: $S \approx 2.1$ meters per second

Math Application by Topic: Piano Pitch
Math Application by Concept: Exponents and Roots
Answers:
a.   Constant of variation, $k \approx 19$
b.   The tension would have to be increased by a multiple of four to double the frequency.

Math Application by Topic: Population Growth
Math Application by Concept: Quadratic Equations
Answer: 1904

Math Application by Topic: Atlatl
Math Application by Concept: Quadratic Equation
Answers:
a. $S = 252$ feet
b.   The arrow would fall back to earth in 12 seconds.

Math Application by Topic: Optical Lab
Math Application by Concept: Quadratic Equations
Answer: $G = 1085$ pairs of glasses

Math Application by Topic: Fighting Fires
Math Application by Concept: Quadratic Equations
Answer: $V = 87$ feet per second

Math Application by Topic: Line of Sight
Math Application by Concept: Quadratic Equations
Answers:
a. $A = \dfrac{D^2}{1.44}$
b.   1,100 feet

Math Application by Topic: Electronics
Math Application by Concept: Quadratic Equations
Answer: $R = -11 - 7i$

Math Application by Topic: Earthquakes
Math Application by Concept: Additional Topics
Answer: The shock wave was approximately 53 times the intensity.

Math Application by Topic: Earthquakes
Math Application by Concept: Additional Topics
Answer: $M = 5$

Math Application by Topic: Piano Pitch
Math Application by Concept: Additional Topics
Answers

a. $\dfrac{1}{6}$

b. $a_{78} = \dfrac{1}{78}$

c. $a_n = \dfrac{1}{n}$

MULTIMEDIA CD-ROM SINGLE USER LICENSE AGREEMENT

1. **NOTICE.** WE ARE WILLING TO LICENSE THE MULTIMEDIA CD-ROM PRODUCT TO YOU ONLY ON THE CONDITION THAT YOU ACCEPT ALL OF THE TERMS CONTAINED IN THIS LICENSE AGREEMENT. PLEASE READ THIS LICENSE AGREEMENT CAREFULLY BEFORE OPENING THE SEALED DISC PACKAGE. BY OPENING THAT PACKAGE YOU AGREE TO BE BOUND BY THE TERMS OF THIS AGREEMENT. IF YOU DO NOT AGREE TO THESE TERMS WE ARE UNWILLING TO LICENSE THE MULTIMEDIA PRODUCT TO YOU, AND YOU SHOULD NOT OPEN THE DISC PACKAGE. IN SUCH CASE, PROMPTLY RETURN THE UNOPENED DISC PACKAGE AND ALL OTHER MATERIAL IN THIS PACKAGE ALONG WITH PROOF OF PAYMENT, TO THE AUTHORIZED-DEALER FROM WHOM YOU OBTAINED IT FOR A FULL REFUND OF THE PRICE YOU PAID.

2. **Ownership and License.** This is a license agreement and NOT an agreement for sale. It permits you to use one copy of the Multimedia Product on a single computer. The Multimedia Product and its contents are owned by us or our licensors, and are protected by U.S. and international copyright laws. Your rights to use the Multimedia Product are specified in this Agreement, and we retain all rights not expressly granted to you in this Agreement.

- You may use one copy of the Multimedia Product for your own internal informational use.
- You may not copy any portion of the Multimedia Product to your computer hard disk, network, or any other media other than printing out or downloading insubstantial portions of the text and images in the Multimedia Product for your own internal informational use.
- You may not copy any of the documentation or other printed materials accompanying the Multimedia Product.

Neither concurrent use on two or more computers nor use in a local area network or other network is *permitted without* separate authorization and *the payment* of additional license fees. *You may* obtain *such authorization by contacting Customer* Service at 1-800-447-9457 and paying the applicable *network license* fee.

3. **Transfer and Other Restrictions.** You may not rent, lend, or lease this Multimedia Product. You may not and you may not permit others to (a) disassemble, decompile, or otherwise derive source code from the software included in the Multimedia Product (the "Software"'), (b) reverse engineer the Software, (c) modify or prepare derivative works of the Multimedia Product, (d) use the Software in an on-line system or (e) use the Multimedia Product in any manner that infringes the intellectual property or other rights of another party.

However, you may transfer this license to use the Multimedia Product to another party on a permanent basis by transferring this copy of the License Agreement, the Multimedia Product, and all documentation. Such transfer of possession terminates your license from us. Such other party shall be licensed under the terms of this Agreement upon its acceptance of this Agreement by its initial use of the Multimedia Product. If you transfer the Multimedia Product, you must remove the installation files from your hard disk and you may not retain any copies of those files for your own use.

4. **Limited Warranty and Limitation of** Liability. For a period of sixty (60) days from the date you acquired the Multimedia Product from us or our authorized dealer, we warrant that the media containing the Multimedia Product will be free from defects that prevent you from installing the Multimedia Product on your computer. If the disc fails to conform to this warranty, you may, as your sole and exclusive remedy, obtain a replacement free of charge if you return the defective disc to us with a dated proof of purchase. Otherwise the Multimedia Product is licensed to you on an "AS IS" basis without any warranty of any nature.

WE DO NOT WARRANT THAT THE MULTIMEDIA PRODUCT WILL MEET YOUR REQUIREMENTS OR THAT ITS OPERATION WILL BE UNINTERRUPTED OR ERROR-FREE. WE EXCLUDE AND EXPRESSLY DISCLAIM ALL EXPRESS AND IMPLIED WARRANTIES NOT STATED HEREIN, INCLUDING THE IMPLIED WARRANTIES OF MERCHANTABILITY AND FITNESS FOR A PARTICULAR PURPOSE.

WE SHALL NOT BE LIABLE FOR ANY DAMAGE OR LOSS OF ANY KIND ARISING OUT OF OR RESULTING FROM YOUR POSSESSION OR USE OF THE MULTIMEDIA PRODUCT (INCLUDING DATA LOSS OR CORRUPTION), REGARDLESS OF WHETHER SUCH LIABILITY IS BASED IN TORT, CONTRACT OR OTHERWISE AND INCLUDING, BUT NOT LIMITED TO ACTUAL, SPECIAL, INDIRECT, INCIDENTAL OR CONSEQUENTIAL DAMAGES. IF THE FOREGOING LIMITATION IS HELD TO BE UNENFORCEABLE, OUR MAXIMUM LIABILITY TO YOU SHALL NOT EXCEED THE AMOUNT OF THE LICENSE FEE PAID BY YOU FOR THE MULTIMEDIA PRODUCT. THE REMEDIES AVAILABLE TO YOU AGAINST US AND THE LICENSORS OF MATERIALS INCLUDED IN THE MULTIMEDIA PRODUCT ARE EXCLUSIVE.

Some states do not allow the limitation or exclusion of implied warranties or liability for incidental or consequential damages, so the above limitations or exclusions may not apply to you.

5. **United States Government Restricted Rights.** The Multimedia Product and documentation are provided with Restricted Rights. Use, duplication or disclosure by the U.S. Government or any agency or instrumentality thereof is subject to restrictions as set forth in subdivision (c)(1)(ii) of the Rights in Technical Data and Computer Multimedia Product clause at 48 C.F.R. 252.227-7013, or in subdivision (c)(1) and (2) of the Commercial Computer Multimedia Product '-Restricted Rights Clause at 48 C.F.R. 52.227-19, as applicable. Manufacturer is Harcourt, Inc., 301 Commerce Street, Fort Worth, TX 76102.

6. **Termination.** This license and your right to use this Multimedia Product automatically terminate if you fail to comply with any provisions of this Agreement, destroy the copy of the Multimedia Product in your possession, or voluntarily return the Multimedia Product to us. Upon termination you will destroy all copies of the Multimedia Product and documentation.

7. **Miscellaneous Provisions.** This Agreement will be governed by and construed in accordance with the substantive laws of the Commonwealth of Pennsylvania. This is the entire agreement between us relating to the Multimedia Product, and supersedes any prior purchase order, communications, advertising or representations concerning the contents of this package. No change or modification of this Agreement will be valid unless it is in writing, and is signed by us.

SYSTEM REQUIREMENTS

For Windows:
Requires IBM or Compatible Pentium Base
200 MHz or higher
Windows 95, 98, or NT Operating System
16 MG of RAM
16 bit sound card
256 color or greater Video Card (High Color or
        greater recommended)
6X CD-ROM Drive
10 MG free Hard Drive space

INSTALLATION INSTRUCTIONS

This CD-ROM requires QuickTime 4.0 for the video playback. If you do not have have it already installed, it is provided on the CD-ROM. If you are not sure if QuickTime 4.0 has already been installed on your computer, it is recommended that you install it. It will not cause any problems if QuickTime 4.0 was previously installed.

WINDOWS 95 or higher
To Install the QuickTime Player (Skip if QuickTime is already installed.)
- Insert the CD-ROM.
- After the CD-ROM auto-starts, click Exit button.
- Return to Windows Desktop.
- Double click on the My Computer icon.
- Double click on the CD-ROM drive icon.
- Double click on the PCFULLQT folder.
- Double click on the QuickTimeInstaller.exe icon.
- Follow the screen prompts.

To run Math In Practice: An Applied Video Companion CD-ROM.
1. Insert the CD-ROM
2. CD-ROM will auto-start.

TECHNICAL SUPPORT

For support call 1-800-447-9457 from 7am to 6pm Central Time Monday through Friday or visit our website at http://www.hbtechsupport.com. Questions and concerns can be faxed to 800-352-1680 or 800-354-1774. You may also email your concerns to tschbc@hbtechsupport.com